SYMMETRY DISCOVERED

CONCEPTS AND APPLICATIONS IN
NATURE AND SCIENCE

SYMMETRY DISCOVERED

CONCEPTS AND APPLICATIONS IN
NATURE AND SCIENCE

JOE ROSEN

TEL-AVIV UNIVERSITY

CAMBRIDGE UNIVERSITY PRESS

CAMBRIDGE

LONDON : NEW YORK : MELBOURNE

Published by the Syndics of the Cambridge University Press
The Pitt Building, Trumpington Street, Cambridge CB2 1RP
Bentley House, 200 Euston Road, London NW1 2DB
32 East 57th Street, New York, NY 10022, USA
296 Beaconsfield Parade, Middle Park, Melbourne 3206, Australia

© Cambridge University Press 1975

Library of Congress catalogue card number: 75–6006

ISBN: 0 521 20695 2

First published 1975

Typeset by E.W.C. Wilkins Ltd., London and Northampton
and printed in Great Britain at the University Printing House, Cambridge
(Euan Phillips, University Printer)

Contents

To the memories of
EMMY NOETHER
and
HERMANN WEYL

Foreword

This foreword is intended for those readers already familiar with the subject. If you are among the uninitiated, however, even though this book was prepared especially for you, I suggest you skip on to the preface.

The book you are about to read was conceived, gestated and published as a result of my discovery of a desert. When I became convinced some time ago that the concepts and the principles of application of symmetry can, and should be, taught and understood much earlier than at the graduate or post-doctoral level, an exploratory expedition into bookland disclosed that between the coastal plain of a couple of children's books and the heights of Weyl's *Symmetry* there was nothing but barren wilderness. So here we are.

How can this attempt to make the desert bloom be described in relation to Weyl's classic? Mainly, it is intended for quite a different audience. While Weyl was addressing the faculty of Princeton University and the Institute of Advanced Study, my 'typical' reader is an advanced high school student or first or second year college student, interested in or specializing in the physical sciences. Additionally, our coverages of material differ in breadth and depth. For example, I discuss the symmetry principle and its application and work out several examples. Weyl does not. On the other hand, he goes into the esthetic aspect of symmetry, which I refrain from doing.

In order to reduce the chance of misunderstanding, I would like to explain here the approaches I chose to use for certain subjects. The most problematic of these is group theory. Now not only is group theory the most suitable mathematical language for symmetry, but the two are actually inseparable, since any group defines a symmetry (an equivalence relation) through its orbits, and, conversely, any symmetry (equivalence relation) defines a group. So group theory should be included in the presentation. Yet not every 'typical' reader is inclined to the abstraction of group theory, and many potential readers might be put off by it. Also, nothing can really be done with group theory at this level. So perhaps it should better be left out. The approach taken here is to include group theory as a distinct 'trail', to segregate all group theory material into clearly marked concentrations. Then he who will, will read everything, and he who won't, will skip the group theory section and all the group theory paragraphs throughout the text.

The way I present geometric symmetry is liable to raise some eyebrows. The discussion is divided, as usual, into separate discussions for 1-, 2- and 3-dimensional systems. But instead of defining the transformations for each kind of system within the dimensionality of the system, geometric transformations are always defined in 3-dimensional space. Thus, for example, reflection of a 1-dimensional system is plane reflection, with the mirror perpendicular to the direction of the system and intersecting it at the reflection center. This approach has, I feel, certain pedagogical advantages. Moreover, it helps emphasize the fact that all physical systems are really 3-dimensional, although for some purposes they might be considered 2- or 1-dimensional.

There are some terms (such as 'lattice') that I chose to use somewhat loosely, and hope that all cases of loose terminology found in this book are among these.

One subject that could usefully be studied after symmetry is approximate symmetry. This should include stability considerations, that is, consideration of the symmetry and approximate symmetry of an effect, given the approximate symmetry of a cause. Here is the den where 'spontaneous symmetry breaking' lurks. However, of all this wealth I chose to include only a brief discussion of the general idea of approximate symmetry, with no mention of general formalism or of stability.

For the 'typical' reader this book should be fairly self-contained. Yet a very large bibliography is included and referred to extensively. My intention is to arouse the reader's curiosity about any of various fields where symmetry plays a role and to tempt him into further reading.

My hopes for this book are that it will fill the niche for which it was designed and that it will encourage other symmetry-minded people to try their hand at rectifying the dearth of published material on symmetry.

Preface

> They expounded the reazles
> For sneezles
> And wheezles,
> The manner of measles
> When new.
> They said 'If he freezles
> In draughts and in breezles,
> Then PHTHEEZLES
> May even ensue.'
> (A.A. Milne: *Now We Are Six*)

Symmetry is like a disease. Or, perhaps more accurately, it *is* a disease. At least in my case; I seem to have a bad case of it. Let me tell you how this came about.

I must always have had a tendency to symmetry. An early mild symptom was a special liking for series of similar things: pads of paper, piles of filing cards, sets of pencils and crayons. My drawings and doodlings inclined to periodicity. Though a lover of serious music, I have long had a special place in my heart for marches (with their strict rhythm). But things took a turn for the worse when I started work toward my Ph.D. degree. I had decided to do my thesis research in the field of what was then called theoretical elementary particle physics. (Since the question of elementarity is an open one, we now prefer to call the field 'high energy physics' or 'physics of particles and fields'.) My thesis adviser introduced me to this field through the study of group theory and its application to elementary particle symmetries, and I finished my Ph.D. with a thesis entitled 'Several aspects of particle symmetries and their origins'.

Then, after its periods of dormancy and incubation, the disease broke out in ever increasing severity. Although I have retained an interest in the physics of particles and fields, most of my research has been devoted to the mathematical and theoretical aspects of symmetry in general and the principles of its application in physics. I have become an avid symmetry fan, addicted beyond cure, utterly convinced of the

ix

fertility of symmetry in scientific study and research as a unifying, clarifying and simplifying factor. Moreover, far from being painful, these severe symptoms afford much pleasure, as I find in those aspects of symmetry with which I am concerned an esthetic enjoyment akin to that brought about by the visual and auditory aspects of symmetry, symmetry in art and music.

And now, as the very writing of this book proves, the disease has reached its contagious stage. I am out looking for converts and will attempt to infect you, the reader, if only slightly, with the symmetry disease. In the following chapters I will introduce you to symmetry and to various fields, mostly in the physical sciences, where symmetry plays a role. A bibliography is appended and referred to often in the text (my method of indicating references is explained there), in the hope that you will, after or while reading this book, follow up by additional reading those aspects of symmetry or symmetry related fields that arouse your interest or curiosity. I have scattered problems throughout the text and hope that you will attempt to work them out, since I think that this helps one understand better. But if you choose to skip them, this should not hamper your reading.

Group theory is the mathematical language of symmetry. A section on group theory is included in chapter 2, and all group theoretical discussions thereafter are confined to clearly marked paragraphs. This material is optional. I recommend attempting to read it as it appears in the text. But if you find it too abstract, don't hesitate to skip it. In that case you might try it again after the first reading.

There is all too little literature on symmetry in general. In fact the only other book I know of is Hermann Weyl's *Symmetry* (SYM, 1). I strongly recommend this book. It is a must for anyone who finds that he has caught the symmetry bug.

I would like to express my gratitude to the following: the Brown University Theoretical High Energy Physics Group, and especially Professor David Feldman, for their hospitality at the time this book was conceived; Professor Nathan Rosen and Professor Gerald E. Tauber for reading an early version of the manuscript and offering helpful suggestions; the staff of the Tel-Aviv University Library of Exact Sciences and Engineering, and especially Ms Tamar Harari, Director, and Ms Celina Sold, for their valuable assistance; the Institute for the History and Philosophy of Science and the Faculty of Exact Sciences of Tel-Aviv University for grants covering technical expenses connected with preparing the book for publication; the Tel-Aviv University Department of Physics and Astronomy for some of the typing; the Israel National Commission for Basic Research for a research grant during the period of which the final version of the manuscript was prepared; Mr Stanley Himmelhoch for his excellent photographic work; Dr Amos Ar, Mr Moshe Ben-Dov, Dr Dan Eisokowitch and Dr Yaakov Friedman for their help; Professor Morton Hamermesh for an enlightening discussion; several anonymous referees for useful comments; the editorial staff of Cambridge University Press for their help and cooperation; and my wife, Dalia, for saving me from starting this book with '*Symmetry* is derived from the Greek. . .' and for her invaluable help in other ways.

May 1975 J.R.

I
Symmetry: what? where? how?

Sometimes he thought sadly to himself, 'Why?' and sometimes
he thought, 'Wherefore?' and sometimes he thought, 'Inasmuch
as which?' – and sometimes he didn't quite know what he *was*
thinking about.
(A.A. Milne: *Winnie-the-Pooh*)

What is symmetry?

Consider a square of definite size, located at a definite position in
space, and having a definite orientation (fig. 1.1). Among all possible
actions that can be performed on this square, there are some that
will leave it in a condition indistinguishable from its original condition
(fig. 1.2). Which are these?

Since the square must remain a square, all changes of shape are
eliminated (fig. 1.3). The square must retain its size, so size changing
is forbidden (fig. 1.4).

The position of the square in space cannot be altered. We must
therefore reject any movement which displaces its center (fig. 1.5).

Let us rotate the square. To preserve position the axis must pass
through the center. If the axis is perpendicular to the plane of the
square, any rotation will leave the square in its original plane, and, of
these, three are objects of our search: rotation by 90°, 180° or 270°
leaves the square in a condition indistinguishable from its original one
(fig. 1.6). (Rotation by 360° also fulfills this requirement, but is equivalent to no rotation at all and can be ignored. Any rotation of more
than 360° is equivalent to one of less than 360°, which can be found by
subtracting off 360° a sufficient number of times.) No other rotation
does this.

Are other rotation axes passing through the center allowable? Most
are not. Those that are lie in the plane of the square. There are four
of them: the two diagonals and the two lines parallel to one or the
other pair of opposite edges. Only a rotation by 180° is acceptable
here, otherwise the square will not be brought back into its original
plane (fig. 1.7).

Fig. 1.1
Square

Fig. 1.2
Square before and after

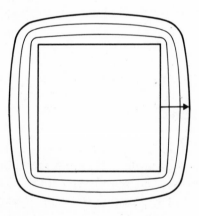

Fig. 1.3
Change shape? No!

Finally, we consider mirror reflections. To keep the center fixed, the plane of the mirror must pass through the center. (The mirror is considered to be two-sided.) It is easily seen that the only mirror orientations that will produce reflections indistinguishable from the original

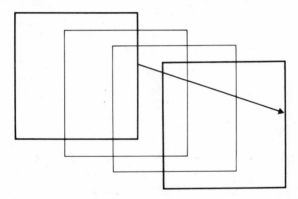

Fig. 1.4
Change size? No!

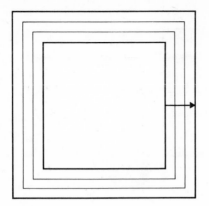

Fig. 1.5
Move? No!

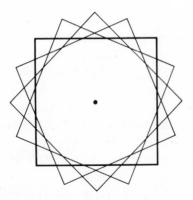

Fig. 1.6
Rotate in plane, about center? Yes, only by 90°, 180°, 270°

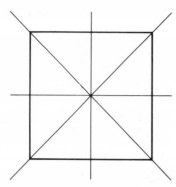

Fig. 1.7
Rotate about other axes? Yes, only about these, by 180°

are: the plane of the mirror is perpendicular to the plane of the square and passes through any one of the four rotation axes of the preceding paragraph (fig. 1.8).

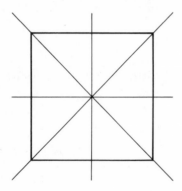

Fig. 1.8
Mirror reflections? Yes, only through these mirrors

This square and all these actions that change it but leave it looking unchanged are an example of *symmetry*. The square is *symmetric* with respect to the actions, which are *symmetry transformations* of the square. The square itself is an example of a *system*. A system is whatever it is that we wish to examine with regard to its symmetry properties. The more different symmetry transformations a system has, the higher its *degree of symmetry*.

Lack of symmetry is *asymmetry*. Our square is asymmetric under change of size, for instance. An example of general asymmetry is the

Fig. 1.9
Asymmetric system

system consisting of our square with the digit 5 drawn on it (fig. 1.9). The addition of 5 certainly does not add new symmetries to the square, and it is easy to see that none of the symmetries that we found for the square remain applicable to the square plus 5. This system is asymmetric.

We emphasize that the system called the 'square' that we investigated so thoroughly is characterized by more than just squareness. Its definition also includes having definite size, position in space, and orientation. If we reduced the strictness of our definition by giving up one or more of these conditions, the degree of symmetry of the system would obviously *increase*. In the extreme, a system characterized solely by its squareness, that is, a square with no additional specifications, is symmetric under everything but change of shape, the only action capable of modifying its single property. In contrast to *our* square, this one is symmetric under change of size, displacement of its center, any rotation about any axis (whether passing through its center or not, or not even passing through the square), and reflection by a mirror having any position and orientation. Various intermediate cases can also be considered, and they will have correspondingly intermediate degrees of symmetry. These examples show how important it is to be fully aware of what is and what is not essential to the system whose symmetry is being considered.

These points will be considered in detail in the following chapters.

Where is symmetry?

What can be a system and have symmetry? Anything can. And not only things. To emphasize the generality of the concept we present a few examples: a system could be a geometric figure, like our square, or a physical entity such as an elementary particle, an atom, a molecule, a crystal, a plant, an animal, the earth, the solar system, our galaxy, or the whole universe. It could be a process taking place in time: the scattering of elementary particles by each other, a chemical reaction, the fall of a stone, a beam of light, biological growth, a piece of music, the flight of men to the moon, the evolution of the solar system, or the development of the universe. The system might even be abstract: the laws of physics, an idea or concept, a mathematical relation, a feeling. In fact, I cannot find a system to which the concept of symmetry is inapplicable.

So where, then, is symmetry? Symmetry can be anywhere!

How is symmetry?

What are the possible symmetry transformations? These can be as diverse and imaginative as the possible systems upon which they might act. The first to come to mind are the geometric transformations considered above: change of shape, change of size, displacement, rotation, reflection. A symmetry transformation might concern time: displacement in time, reversal of chronological order of events, change of size of time intervals. A system might be symmetric with respect to interchange of its parts. Physicists work with symmetry under transformations such as interchange of positive and negative electric charge, interchange of particles and antiparticles, change of velocity (this has to do with special relativity), and various abstract symmetry transformations best defined in the context of the physical system to which they are relevant.

These and other examples of symmetry transformations and systems on which they act are presented and discussed in later chapters.

2
The language of symmetry

Concepts and terminology

'Well,' said Owl, 'the customary procedure in
such cases is as follows.'
    'What does Crustimoney Proseedcake mean?'
said Pooh. 'For I am a Bear of Very Little Brain,
and long words Bother me.'
(A.A. Milne: *Winnie-the-Pooh*)

After making a brief initial acquaintance with the concept of symmetry
and its general applicability, we now examine the concepts more
closely and familiarize ourselves with some important terms, which
we need for our discussions.

A *system*, as mentioned in the previous chapter, is any object of
interest with regard to its symmetry properties. The examples of the
previous chapter illustrate the generality of the concept of a system;
it can be abstract or concrete, microscopic or macroscopic, static or
dynamic, finite or infinite.

A possible condition of a system is referred to as a *state*. For example,
the system of the square that we investigated in some detail in the
previous chapter has many (actually an infinite number of) states.
Each state is determined by specifying the size of the square, its location
in space, and its orientation. In contrast, the system whose sole
characteristic is squareness, so that properties such as size, location
and orientation are not relevant to it, has only one state, since there
are not various degrees of squareness.

(Concerning this latter system, the following discussion might be
of interest. It is valid to claim that the system has two states, squareness and nonsquareness. This is because we consider the possibility of
changing its shape (but see that this is obviously not a symmetry
transformation), and how can such a possibility be considered, unless
it has a state of nonsquareness? We could even generalize and allow
the system to have an infinite number of states – all possible shapes.
We could then define the system as 'a geometric figure having a definite shape'; it would not have to be specifically a square. The symmetry

of this system is the same as that of the system characterized solely by squareness: any transformation that does not change shape is a symmetry transformation. Therefore, the *symmetry* of a system characterized solely by shape is symmetric under change of shape! If that was too twisted a piece of reasoning, just ignore it and continue.)

We have used the term *transformation* without really defining it. It is common to think of a transformation as an action that changes a system from some initial state to some other final state. This is not wrong, but it is too limited a definition for our purpose.

First, we would prefer reducing any emphasis on the action that causes the change of the system while increasing the importance of the relationship 'initial state → final state'. The final state, called the *image*, should be considered as derived from the initial state and related to it. We should think less of performing an actual action, which changes the system from the initial state to the image state, and more of setting up a correspondence – to the initial state of the system we make correspond another state as its image.

Second, such a correspondence should be set up, not for just a single state of the system, but for all states. A transformation acts on a system no matter what state it is in. So a transformation is a rule of some kind, whereby the appropriate image can be derived for every state of the system. Any 'action' involved might easily differ according to which state the system happens to be in, and indeed it is hardly reasonable to think in terms of actually performing 'actions' for all states simultaneously. The correspondence picture is by far the most suitable, and this is the meaning we shall reserve for transformation.

The rule of correspondence, by which an image state is associated with every state of a system, is almost completely arbitrary, limited almost solely by the imagination of the inventor of the transformation. (Of course, some transformations are more useful than others.) This correspondence might be expressed either as a general prescription, in which the relation between every state and its image is described in general terms (fig. 2.1), or it might be exhibited as a double list, like a translating dictionary, with all states of the system listed in the first column and their corresponding images listed respectively across from them in the second column (fig. 2.2). For a system with an infinite number of states, the general prescription is the only possible one. The only limitation on this correspondence is that every state of the system must appear in the set of image states and must appear only once, so that the images of different states are always different. This is called

Fig. 2.1
Transformation by general prescription

Initial state of system	Image state of system
A	G
B	W
C	D
D	A
E	B
⋮	⋮

Fig. 2.2
Transformation by double list

a *one-to-one* correspondence and completes the definition of a transformation.

Returning to our square, for example, we have a system with an infinite number of states, and any transformation must be described in general terms. The transformation 'rotation by 90° about the axis through its center and perpendicular to its plane' is just such a prescription, even though it is formulated in terms of an 'action'. As a 'state → image' correspondence this transformation means that, whatever the size of the square, wherever it is located in space, whatever the orientation of the plane in which it lies, and whatever its own orientation in that plane, the appropriate image has the same size and location in space and lies in the same plane, but its orientation in this plane differs from that of the original state by an angle of 90°.

To show an example of the double list kind of transformation, we need a system with a finite number of states. Let this consist of three depressions in the sand and a ball lying in any one of them. The system has three states, which we label *A*, *B*, *C* according to which depression contains the ball. A transformation for this system is a specification of where the ball is to be finally placed if it is initially found in each possible depression. The following is a possible transformation:

Initial state (depression)	Image state (depression)
A	B
B	C
C	A

Another possible transformation is:

Initial state (depression)	Image state (depression)
A	B
B	A
C	C

Another possibility is:

Initial state (depression)	Image state (depression)
A	C
B	B
C	A

Other transformations are easily set up.

In brief then, a transformation is a one-to-one assignment of an image state to each state of a system.

A transformation that completely cancels the effect of another transformation is called the *inverse* of the latter. The inverse transformation acts as follows. A transformation assigns an image state to every state of the system. Its inverse transformation also does this, but in general makes a different assignment, such that the image which it assigns to each image of the original transformation is just the corresponding initial state (fig. 2.3).

The inverse of the transformation of rotation by 90° about a given axis, for example, is rotation by an additional 270° about the same axis, producing a total rotation of 360° which is no rotation at all (fig. 2.4). (The transformation of rotation by 90° the other way is also a valid inverse, but it is equivalent to rotation by 270° and so can be ignored.)

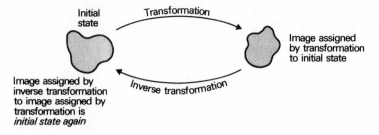

Fig. 2.3
Inverse of transformation

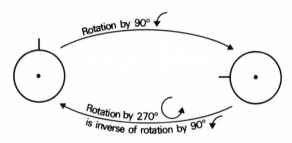

Fig. 2.4
Example of inverse of transformation

The inverse of the transformation

$$\begin{pmatrix} A \\ B \\ C \end{pmatrix} \longrightarrow \begin{pmatrix} B \\ C \\ A \end{pmatrix}$$

for the system of three depressions and a ball is the transformation

$$\begin{pmatrix} A \\ B \\ C \end{pmatrix} \longrightarrow \begin{pmatrix} C \\ A \\ B \end{pmatrix}$$

If the system is initially in state A, the original transformation puts it in state B. The inverse then returns it to state A. If it is initially in state B, it is transformed to C, then back to B. And similarly, from C to A, then back to C.

The inverse of the transformation

$$\begin{pmatrix} A \\ B \\ C \end{pmatrix} \longrightarrow \begin{pmatrix} B \\ A \\ C \end{pmatrix}$$

happens to be just this same transformation itself. If the system is in state A, the transformation puts it in state B. Another application of the transformation returns it to state A. State B becomes A, then turns back into state B. State C remains state C in both steps.

So the inverse of any transformation is found simply by reversing the arrow of the correspondence.

When it happens that a transformation affects a system in such a way that all images are indistinguishable from their respective initial states, the system is said to be *invariant* or *symmetric* under the transformation. This transformation is then called an *invariance transformation* or a *symmetry transformation* of the system (fig. 2.5). A square, for example, is invariant under rotation by 90° about an axis through its center and perpendicular to its plane, since its shape is such that the result of this rotation is indistinguishable from the initial state for every possible initial state.

The general term *symmetry* means invariance under one or more transformations. The more different transformations a system is invariant under, the higher its *degree of symmetry*.

A system is *asymmetric* under a transformation if it is not invariant under the transformation, and is completely asymmetric if it has no symmetry transformations at all. For example, a square with the digit 5 added (chapter 1) is asymmetric under rotations and reflections.

The set of all symmetry transformations of a system comprises the *symmetry group* of the system. The term *group* is used here in its precise mathematical sense and implies that this set has certain very definite properties. For the interested reader a brief introduction to group

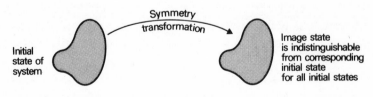

Fig. 2.5
Symmetry transformation

theory is presented in the next section of this chapter. At this point
we shall only look at some of the more important (for the purpose of
our discussion) features of the symmetry group. Some references are
Weyl (SYM, 1), Bell and Fletcher (GRP, 4), Bell (GRP, 3), Linn
(GEO, 2), and Shubnikov (COL, 1).

Most worthy of note is the property that the transformation consisting of the consecutive application of two symmetry transformations,
the second acting on the image of the first, is also a symmetry transformation. This is easily seen as follows. Denote the initial state of
the system by A. The first symmetry transformation transforms it to
some state B, which is indistinguishable from A. The second symmetry
transformation transforms state B into some state C, which is indistinguishable from B and therefore also from A. The combined transformation brings the system from state A to state C, and, since these
states are indistinguishable, it is a symmetry transformation (fig. 2.6).

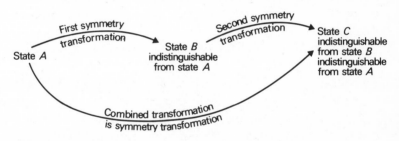

Fig. 2.6
Combinations of symmetry transformations are also symmetry transformations

It then follows that the transformation consisting of the consecutive
application of any number of (the same or different) symmetry transformations is again a symmetry transformation. As an example consider our square of the previous chapter and let all rotations take place
about the axis through its center and perpendicular to its plane. Assume
we have discovered that rotation by $90°$ is a symmetry transformation.
Let us perform a rotation by $90°$ followed by another rotation by $90°$.
This is equivalent to a rotation by $180°$, and according to the group
property it must be a symmetry transformation. Indeed, we see that
it is. Now apply a rotation by $180°$ followed by a rotation by $90°$.
This is equivalent to a total rotation by $270°$, which again must be a
symmetry transformation, as in fact it is (fig. 2.7).

The square is a very simple example, of course, and it is quite easy

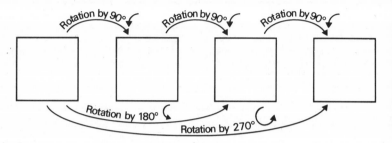

Fig. 2.7
Rotations by multiples of 90° are symmetry transformations of square

to see that rotations by 90°, 180° and 270° are symmetry transformations. The point we would like to emphasize here is that for any system whatsoever, given one or more but not all of its symmetry transformations, it is often possible to discover more symmetry transformations by examining the repeated application of each known symmetry
transformation and the consecutive application of various combinations of them.

Another feature of the symmetry group is that the inverse of every
symmetry transformation is also a symmetry transformation. In addition, the inverse of the inverse of a symmetry transformation is, of
course, the symmetry transformation itself (fig. 2.8). In the case of
our square, for example, we see that the inverse of the symmetry transformation of rotation by 90° is rotation by 270°, which is indeed a

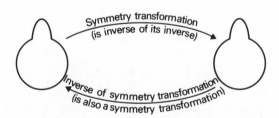

Fig. 2.8
Inverse of symmetry transformation is also symmetry transformation

symmetry transformation. And the inverse of the inverse here is the
inverse of rotation by 270°, which is seen to be rotation by 90°, as it
should be (fig. 2.9).

An important fact of transformations in general, and not just of

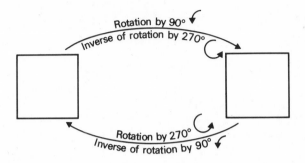

Fig. 2.9
Rotations by 90° and 270° are mutually inverse symmetry transformations of square

symmetry transformations, is that the result of consecutive application of two transformations might depend on their order. It is clear that this is not the case for rotations about a common axis. Here the transformation consisting of rotation by angle A followed by rotation by angle B is identical with the transformation of rotation by angle B followed by rotation by angle A. They are both equivalent to rotation by angle $A + B$ (fig. 2.10). Transformations of this type, whose order is immaterial, are said to *commute*.

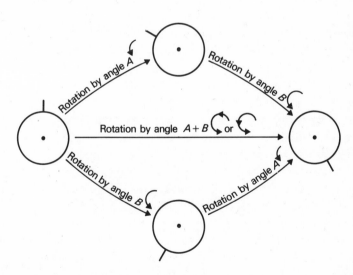

Fig. 2.10
Example of commuting transformations

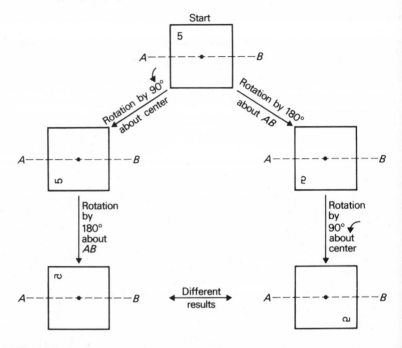

Fig. 2.11
Example of noncommuting transformations

Rotations about different axes, however, do not in general commute. As an example consider the marked square in fig. 2.11. We apply the following two transformations consecutively in each possible order: rotation by 90° about the axis through its center and perpendicular to its plane and rotation by 180° about axis *AB* (lying in the plane of the square, passing through its center, and parallel to a pair of edges). The two results are different, as indicated by the final positions of the digit 5. Pairs of transformations like these, whose order makes a difference, are called *noncommuting.*

It is simple to construct the inverse of a transformation composed of the consecutive application of noncommuting transformations. This is just the consecutive application of the inverses of these same transformations, but *in reverse order* (fig. 2.12).

As an example of this general prescription, examine the transformation taking us from an empty uncovered cooking pot to a full covered one (fig. 2.13). This involves two consecutive noncommuting transformations: (1) pouring water into the pot, and (2) putting the lid on. (Confirm the noncommutativity of these transformations by imagining

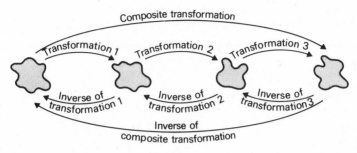

Fig. 2.12
Inverse of composite transformation

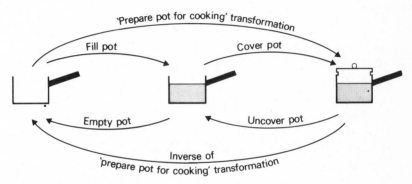

Fig. 2.13
Example of inverse of composite transformation

the result of performing them in opposite order.) The inverse of this transformation consists of: (1) taking off the lid (the inverse of putting it on), and (2) pouring out the water (the inverse of pouring water in). Note the reversal of order, which seems so natural in the example. Consider what would be the result of consecutively performing the inverses without reversing the order.

The inverse of a transformation consisting of consecutive application of commuting transformations is even simpler. It is the consecutive application of the inverses in any order at all.

The language of symmetry

Group theory (optional)

'Let's have a look,' said Eeyore, and he turned slowly round
to the place where his tail had been a little while ago, and then,
finding that he couldn't catch it up, he turned round the other
way, until he came back to where he was at first. . .
(A.A. Milne: *Winnie-the-Pooh*)

The theory of groups is a perfectly respectable branch of mathematics, even though to many people it might not be as familiar as arithmetic, geometry or calculus, and to many more it might not be familiar at all. But just as arithmetic is the most suitable language for financial transactions and geometry is most appropriate for investigating the properties of spatial figures and bodies, so is group theory best suited to the study of symmetry. This is due to the fact, as we mentioned in the previous section, that the set of all symmetry transformations of a system forms a group. An understanding of group theory is thus an invaluable companion in the quest for insight into symmetry and is indispensible for serious study of the subject. The scope of this book, though, does not allow more than a very brief introduction to group theory, and no application of the theory is attempted. If the reader finds his interest aroused and would like to look further into group theory, the bibliography contains suggestions for additional and more advanced reading listed under GRP, including cross-references.

We start by defining an *abstract group*, for which we use the symbol *G*. This is, first of all, a *set* (= collection) of *elements*, which we denote by *a, b, c, . . .* This set must be endowed with a *law of composition*, which is a double way of combining any two elements of the set. If *a* and *b* are elements of the set, then their compositions are written *ab* and *ba*. Now to be a group, this set together with its law of composition must satisfy the following conditions:

(1) For any two elements of the set *a, b* both *ab* and *ba* are also elements of the set. This is called *closure*.

(2) The composition is *associative*; that is, $a(bc) = (ab)c$. In other words, in the composition of three elements the order of combining pairs is immaterial. Thus one can evaluate *abc* by first making the composition $bc = d$ and then forming *ad*, corresponding to $a(bc)$, or one can start with $ab = f$ and then form *fc*, corresponding to $(ab)c$. The results must be identical. Then this will hold as well for compositions of more than three elements.

(3) The set contains an element *e* called the *identity*, such that for every element *a* of the set $ea = ae = a$. The characteristic property of

the identity is, therefore, that its composition in either order with any element of the set is just that element itself.

(4) For every element a of the set there is an element of the set called the *inverse* of a, which we denote a^{-1}, such that $aa^{-1} = a^{-1}a = e$. In other words, for every element of the set there is an element whose composition with it in either order gives the identity.

If the composition of two elements a, b of a group does not depend on their order, that is, $ab = ba$, then these elements are said to *commute*. From the definition of a group it is seen that the identity commutes with all elements of the group and that every element commutes with its inverse. Obviously every element of a group commutes with itself. If all the elements of a group G commute with each other, that is, $ab = ba$ for *all* a, b which are elements of G, the group is called *commutative*. Otherwise it is *noncommutative*.

The number of elements in a group is called its *order* and can be finite or infinite.

As an example of group theoretical proof, let us show that the identity is unique, that is, there is only one such element. (Note that condition 3 only requires that such an element exists, but does not exclude that more than one exists.) We start by assuming that it is not unique, so that there is more than one element obeying the identity definition. Denote any two to these by e', e''. Then $e'a = ae' = e''a = ae'' = a$, where a is any element of the group. Evaluate the relation $e'a = a$ for the specific element $a = e''$. We get $e'e'' = e''$. Now evaluate $ae'' = a$ for $a = e'$. We get $e'e'' = e'$. Comparing these results, we obtain $e' = e''$, which is in contradiction to our assumption that e' and e'' are different. The identity is therefore proved to be unique.

The inverse of an element is unique; that is, for every element a of G there is only one element, which we denoted a^{-1}, such that $aa^{-1} = a^{-1}a = e$. (Note that condition 4 only requires that such an element exists, not that it is unique.) As an additional example of abstract group theoretical calculation, we present a proof of this statement. We do this by showing the impossibility of the opposite assumption. So assume that the element a has more than one inverse, and denote any two of them by b, c. Then $ab = ba = ac = ca = e$. According to the associativity condition, $c(ab) = (ca)b$. The left hand side equals $c(ab) = ce = c$, while the right hand side is $(ca)b = eb = b$. Therefore $b = c$, and the result is that, even if we assume that element a has different inverses, we find that they cannot be different. This is a contradiction and proves the impossibility of an element having more than one inverse.

From the definition of the inverse and its uniqueness it is clear that the inverse of the inverse is the original element itself; that is, the inverse of a^{-1} is a. This can be written $(a^{-1})^{-1} = a$.

The inverse of a composition of elements is the composition in opposite order of the inverses of the individual elements. In symbols this means that the inverse of ab is $b^{-1}a^{-1}$, or $(ab)^{-1} = b^{-1}a^{-1}$. This is verified directly as follows:

$$
\begin{aligned}
(ab)(b^{-1}a^{-1}) &= a(bb^{-1})a^{-1} && \text{(associativity)} \\
&= aea^{-1} && \text{(inverse)} \\
&= (ae)a^{-1} && \text{(associativity)} \\
&= aa^{-1} && \text{(identity)} \\
&= e && \text{(inverse)}
\end{aligned}
$$

The verification of $(b^{-1}a^{-1})(ab) = e$ is done similarly. This can be shown to hold as well for compositions of more than two elements, for example, $(abc)^{-1} = c^{-1}b^{-1}a^{-1}$.

The statement of the results of all possible compositions of pairs of elements of a group displays the group's *structure*. For finite order groups this is most clearly done by setting up a *group table* (fig. 2.14),

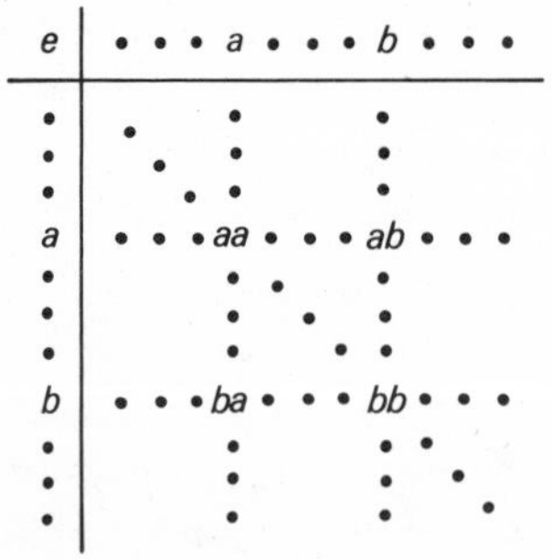

Fig. 2.14
Group table

which is similar to an ordinary multiplication table. To find ab one looks up a in the left column and b in the top row; the composition ab is then found at the intersection of the row starting with a and the column headed by b. For the composition ba it is b that is on the left and a at the top.

When the symbols for the group elements have the same ordering in the top row (left to right) as in the left column (up to down), a group table will be symmetric under reflection through the diagonal if and only if the group is commutative. Note that by reordering the rows or columns or by changing the symbols denoting the group elements a group table can be made to look quite different while still describing the same group. So two seemingly different group tables belong to the same group, if they can be made identical by reordering rows and columns and relabeling elements.

A group table is clearly not possible for groups of infinite order. Instead, one must express the results of all possible compositions by a general rule.

A concrete example of an abstract group, that is, a group of concrete elements having the same structure as an abstract group, is called a *realization* of this abstract group. Group realizations can be found practically everywhere, but examples are usually taken from mathematics, so that they will be as familiar as possible. Such realizations might be groups of numbers, groups of matrices, groups of rotations, or groups of other geometric transformations.

Two groups that are realizations of the same abstract group have the same structure and are called *isomorphic* with each other. Each one is therefore also isomorphic with the abstract group of which they are realizations.

When part of the elements of a group by themselves form a group with the same law of composition, the latter group is called a *subgroup* of the former. Since the identity element by itself forms a group of order 1, and every group by definition has an identity element, all groups therefore contain at least one subgroup.

The set of all symmetry transformations of a system forms a group, the *symmetry group* of the system, if we include as a symmetry transformation the 'transformation' of doing nothing at all. Like any other transformation this one can also be given a 'state → image' interpretation; the image of any state is just that state itself. So even though it does not transform anything in the intuitive sense, the non-action of doing nothing is still considered a transformation, according to our more general definition of the previous section, and serves as the identity element of the symmetry group. It is call the *identity transformation* (fig. 2.15).

Every system is symmetric with respect to the identity transformation. So every system has a symmetry group, even if it is only the trivial group of order 1 consisting of just the identity element. It hardly

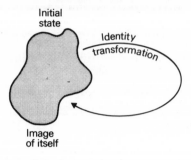

Fig. 2.15
Identity transformation

seems justified to say that a system has symmetry if its only symmetry transformation is the identity transformation, as indeed we must, according to our definition of symmetry in the previous section. We therefore reformulate the definition as follows: A system is said to have symmetry if its symmetry group is of order greater than 1. The larger the order, the higher its degree of symmetry. It is completely asymmetric if its symmetry group is of order 1, that is, consists only of the identity transformation.

PROBLEM

Formulate precisely in group theoretical terms the following properties of the symmetry group of a system, which were loosely described in the previous section:

(*a*) The composition of two symmetry transformations is also a symmetry transformation.

(*b*) The composition of any number of symmetry transformations is also a symmetry transformation.

(*c*) The inverse of a symmetry transformation is also a symmetry transformation.

(*d*) The inverse of a composite transformation is the composition in reverse order of the individual inverses.

(*e*) The inverse of a composition of commuting transformations is the composition in any order of the individual inverses.

PROBLEM

Prove that the set of symmetry transformations of a system (including the identity transformation) indeed obeys the four conditions defining a group and thus forms a group, the symmetry group of the system.

It is an easy matter to construct group tables for all low order abstract groups. We display these for all groups of orders 1 to 5 and one group of order 6 and present one or more examples of realizations for each case but the last.

Order 1. This is the trivial group consisting only of the identity element e. One realization of this group is the number 1 and ordinary multiplication as the composition. Another realization is the number 0 and the composition of ordinary addition. As mentioned above, this group can also be realized by the identity transformation of any system.

Order 2. There is only one group of order 2. It consists of the identity e and one other element a, which must then be its own inverse, $aa = e$. The group table is shown in fig. 2.16. A realization of this group is the set of numbers 1, -1 under multiplication. The number 1 serves as the identity, while -1 is its own inverse, $(-1) \times (-1) = 1$.

$$
\begin{array}{c|cc}
e & a \\
\hline
a & e
\end{array}
$$

Fig. 2.16
Group of order 2

Another, geometrical, realization is the set consisting of the identity transformation and the transformation of mirror reflection with the composition of consecutive reflection. The identity transformation, of course, serves as the group identity element, and the reflection transformation is its own inverse, since two consecutive reflections in the same mirror bring the situation back to its initial state. In a similar manner we could use the isomorphic group of rotations about a common axis by $0°$ and $180°$.

Order 3. There is only one group of this order, and it is commutative. Its table is presented in fig. 2.17. The elements a, b are each the

$$
\begin{array}{c|cc}
e & a & b \\
\hline
a & b & e \\
b & e & a
\end{array}
$$

Fig. 2.17
Group of order 3

inverse of the other, $ab = ba = e$, and the composition of each with itself gives the other, $aa = b$ and $bb = a$. As a realization we can take the rotational symmetry group of the equilateral triangle, which is the set of rotations about a common axis (through the center of the triangle and perpendicular to its plane) by $0°$ (the identity transformation), $120° = 360°/3$, and $240° = 2 \times 120°$, with consecutive rotation as the group composition. The identity transformation of rotation by $0°$ naturally corresponds to the identity element, while the other two rotations can respectively correspond to either a, b or b, a.

Order 4. There are two different abstract groups of order 4. They are both commutative. One of them has a structure similar to that of the group or order 3. Its table is shown in fig. 2.18. It can be realized by the rotational symmetry group of the square, the group of rotations about a common axis (through the center of the square and perpendicular to its plane) by $0°$ (the identity transformation), $90° = 360°/4$, $180° = 2 \times 90°$, and $270° = 3 \times 90°$, corresponding respectively to e, a, b, c. The other group of order 4 has the group table of fig. 2.19.

$$
\begin{array}{c|ccc}
e & a & b & c \\
\hline
a & b & c & e \\
b & c & e & a \\
c & e & a & b \\
\end{array}
$$

Fig. 2.18
First group of order 4

$$
\begin{array}{c|ccc}
e & a & b & c \\
\hline
a & e & c & b \\
b & c & e & a \\
c & b & a & e \\
\end{array}
$$

Fig. 2.19
Second group of order 4

In this group each element is its own inverse, $aa = bb = cc = e$. To obtain a realization we imagine two perpendicular intersecting mirrors A and B and their line of intersection C as in fig. 2.20. Results of the transformations of reflection through each mirror and rotation by $180°$ about their line of intersection are shown in cross-section in fig. 2.21. Each of these transformations is its own inverse. The set of trans-

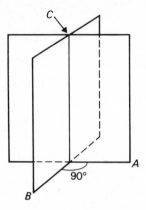

Fig. 2.20
Two perpendicular intersecting mirrors A and B and their line of intersection C

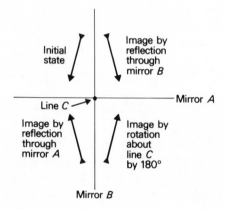

Fig. 2.21
Reflections in perpendicular mirrors A and B and rotation by $180°$ about their line of intersection C, shown in cross-section

formations consisting of these three transformations and the identity transformation, with consecutive transformation as composition, forms a group and is a realization of the second abstract group or order 4.

Order 5. There is only one abstract group of order 5, and it is commutative. Its structure is similar to that of the first group of order 4, and is described by the group table of fig. 2.22. The rotational symmetry group of the regular pentagon is a realization of this abstract group. It is the set of rotations about a common axis (through the center of the pentagon and perpendicular to its plane) by $0°, 72° =$

$$
\begin{array}{c|cccc}
e & a & b & c & d \\
\hline
a & b & c & d & e \\
b & c & d & e & a \\
c & d & e & a & b \\
d & e & a & b & c \\
\end{array}
$$

Fig. 2.22
Group of order 5

$$
\begin{array}{c|ccccc}
e & a & b & c & d & f \\
\hline
a & b & e & f & c & d \\
b & e & a & d & f & c \\
c & d & f & e & a & b \\
d & f & c & b & e & a \\
f & c & d & a & b & e \\
\end{array}
$$

Fig. 2.23
Noncommutative group of order 6

$360°/5$, $144° = 2 \times 72°$, $216° = 3 \times 72°$, and $288° = 4 \times 72°$, with consecutive rotation as the group composition.

All the groups we have encountered so far have been commutative. To find noncommutative groups we must go to higher orders. The lowest order noncommutative group is one of the order 6 groups. To dispel any impression that all groups are commutative, we display its group table in fig. 2.23.

Infinite order groups are very important. We present several examples of these.

(1) The group of all integers (positive, negative and zero) under addition. The identity is 0. The inverse of a number a is its negative $-a$. This group is commutative.

(2) The group of all rational numbers (numbers expressible as a ratio of integers), excluding zero, under multiplication. The identity here is 1. The inverse of a number a is its reciprocal $1/a$. The group is commutative.

(3) The group of all displacements along a common line. The group composition is consecutive displacement. The identity is the null displacement. The inverse of a displacement is a displacement the other

way by the same amount. This group is commutative. It is a subgroup of the symmetry group of an infinite straight line. (The full symmetry group also includes reflections.)

(4) The group of rotations about a common axis. The composition is consecutive rotation. Rotation by $0°$ acts as the identity. The inverse of rotation by $a°$ is rotation by $(360 - a)°$. The group is commutative. It is a subgroup of the symmetry group of a circle. (The full symmetry group also includes reflections.)

(5) The group of rotations about a point. This group consists of all rotations about all axes passing through a fixed point. The composition is again consecutive rotation. This group is noncommutative, as we showed in the example of noncommuting rotations in the previous section. It has as a subgroup the group of rotations about a common axis (the previous example). This group is itself a subgroup of the symmetry group of a sphere. (The full symmetry group also includes reflections.)

3
Geometric symmetry

Geometric symmetry is symmetry with respect to geometric transformations. A *geometric transformation* is any transformation that affects only the geometric properties of a system. This is the kind of symmetry that is usually alluded to, when the word 'symmetry' is mentioned in everyday conversation. It is then related to features such as balance, regularity, order, repetition, proportionality, harmony and similarity. As we emphasized in the first chapter, the concept of symmetry is much broader and more general than just geometric symmetry. But geometric symmetry is obviously important, and it is well worth our while to increase our understanding of it. This is what we attempt to accomplish in the present chapter. References are listed under GEO in the bibliography, including cross-references.

The space with which we are familiar, in which we live, objects exist and events occur, is 3-dimensional. All real systems are 3-dimensional, and geometric transformations can affect them 3-dimensionally. Abstract systems having geometric properties, though, do not have to be 3-dimensional, but can be 2- or 1-dimensional. (As long as they are abstract, they can even be 4-dimensional or n-dimensional, but we do not concern ourselves with these here.) The square of chapter 1, for example, has no thickness and so is a 2-dimensional system. And even real systems might be considered as being 2- or 1-dimensional, when one or two of their dimensions can be ignored for the purposes of the discussion. For instance we can treat a length of pipe as a 1-dimensional system, if we are only interested in symmetries related to its longitudinal dimension. But we can also think of the same length of pipe as a 2-dimensional system, if we are only investigating the shape of its cross-section. It is convenient to consider separately the geometric symmetries of 1-, 2- and 3-dimensional systems. We refer to these respectively as *linear*, *planar* and *spatial* symmetry.

Linear symmetry

Winnie-the-Pooh read the two notices very carefully, first
from left to right, and afterwards, in case he had missed some
of it, from right to left.
(A.A. Milne: *Winnie-the-Pooh*)

The geometric transformations that we discuss for 1-dimensional systems are displacement and plane reflection.

The transformation of *displacement* (or *translation*) involves displacing the system by a given interval in the only direction that the system is considered to have (fig. 3.1). A system can be symmetric under displacement only if it is of infinite extent and so has no ends.

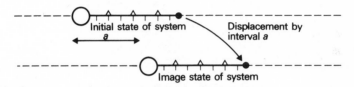

Fig. 3.1
Displacement transformation of 1-dimensional system

An infinite straight line is symmetric under displacements by any interval in the direction of the line (fig. 3.2(*a*)). So too is an infinite straight rod of constant diameter (fig. 3.2(*b*)). Any system that has an

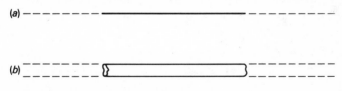

Fig. 3.2
Displacement symmetric 1-dimensional systems

infinite regular repetition in one direction is symmetric under displacements in that direction, but only under those displacements by a certain *minimum displacement interval* and multiples of it. The patterns of fig. 3.3 are examples of such systems. The minimum displacement interval is denoted by *a* in each case.

No real system is infinite, however, so no real system can have exact displacement symmetry. But a finite system is said to have *approximate*

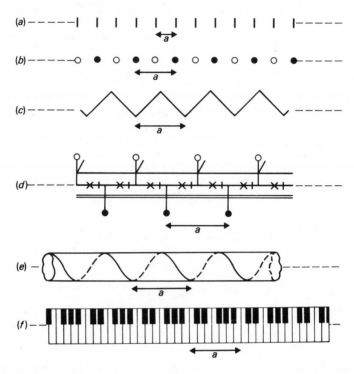

Fig. 3.3
Displacement symmetric 1-dimensional systems with minimum displacement intervals indicated by *a*

displacement symmetry, if it can be imagined as being part of an infinite exactly symmetric system and if the minimum displacement interval is much smaller than the total length of the system. Then we can ignore the effect of the ends, as long as we observe the system sufficiently far from the ends and restrict the displacements to sufficiently small multiples of the minimum displacement interval.

Systems with approximate linear displacement symmetry occur in everyday situations. Some examples are: a long traffic jam, a long train, railroad tracks and their cross-ties, a large pile of data processing cards, telephone poles along a highway, cracks (the ones put in on purpose) along a sidewalk.

Group theory. The set of all displacements along a line, including the identity transformation, with the composition of consecutive displacement, forms a commutative group of infinite order, called the *continuous linear displacement group.* Among these transformations any set consisting of the identity transformation and displacements by all multiples of any minimum displacement interval form a subgroup, also of infinite order, a *discrete linear displacement group*. All such subgroups are isomorphic.

The *plane reflection* (or *plane inversion*) transformation is accomplished by imagining a mirror with its plane perpendicular to the direction of the system and reflecting the system in it. The intersection of the mirror with the line of the system is called the *reflection center*. The image of a point falls therefore on the line of the system, on the opposite side of the reflection center from the initial point, and at the same distance from the reflection center as is the initial point (fig. 3.4). Reflection transformations through different centers do not commute. Both finite and infinite systems can have reflection symmetry.

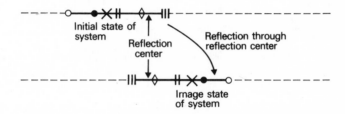

Fig. 3.4
Plane reflection transformation of 1-dimensional system

A line segment has reflection symmetry through a center located at its middle (fig. 3.5(*a*)). An infinite straight line is symmetric under reflection through any center located anywhere along the line (fig. 3.2(*a*)). Fig. 3.5 presents examples of reflection symmetric patterns. Centers of reflection symmetry are indicated by arrows. The systems of fig. 3.3(*d*), (*e*) do not have reflection symmetry, while those of fig. 3.3(*a*)–(*c*), (*f*) do.

PROBLEM

Find the centers of reflection symmetry for the reflection symmetric patterns of fig. 3.3.

Note that if a displacement symmetric system also has plane reflection symmetry, then it has an infinite number of centers of reflection symmetry separated by half the minimum displacement interval. This is shown in fig. 3.5(*d*), (*e*).

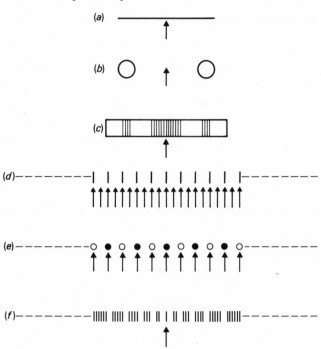

Fig. 3.5
Plane reflection symmetric 1-dimensional systems with centers of reflection symmetry indicated by arrows

PROBLEM
Why is this so?

A common example of a 1-dimensional system with plane reflection symmetry is a loaf of bread.

Group theory. The identity transformation and the transformation of reflection through a given reflection center, with the composition of consecutive reflection, form a group of order 2.

Planar symmetry

You remember how he discovered the North Pole; well, he was so proud of this that he asked Christopher Robin if there were any other Poles such as a Bear of Little Brain might discover.

'There's a South Pole,' said Christopher Robin, 'and I expect there's an East Pole and a West Pole, though people don't like talking about them.'

(A.A. Milne: *Winnie-the-Pooh*)

The geometric transformations of displacement and plane reflection that we considered for 1-dimensional systems are certainly applicable to 2-dimensional systems also. However, the extra dimension makes it possible to apply additional kinds of transformations that are inapplicable to 1-dimensional systems. Of these we discuss glide, rotation and line reflection.

The displacement transformation acts on 2-dimensional systems in the same way that it acts on 1-dimensional systems. But in the present case it is not confined to a single direction and can act in any direction parallel to the plane of the system. All displacement transformations commute, whether they are in the same direction or not (fig. 3.6).

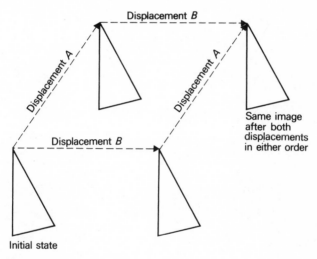

Fig. 3.6
Commutativity of displacement transformations in plane

Exact displacement symmetry is possible only for systems which have infinite extent (no ends) in the direction or directions of displacement. Systems which are homogeneous in these directions are then symmetric under displacements by any arbitrary interval. Systems having infinite regular repetition in one or more directions are symmetric under displacements in those directions by a minimum interval (possibly different for different directions) and multiples of it.

An abstract plane is symmetric under displacement by any interval in any direction parallel to itself. That region of a plane bounded by two parallel straight lines is symmetric under displacements by any

Fig. 3.7
2-dimensional system that is displacement symmetric in single direction

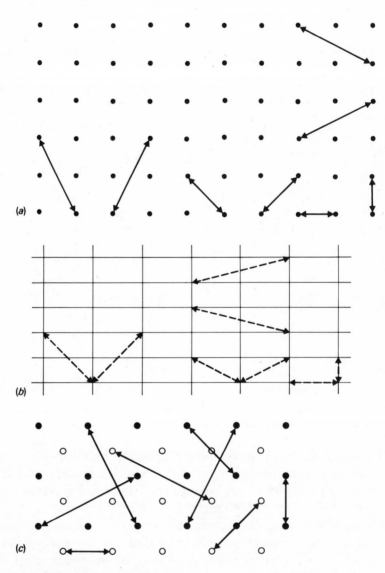

interval in one direction only (fig. 3.7).

A 2-dimensional system that has infinite regular repetition, and
therefore displacement symmetry, in more than one direction is called

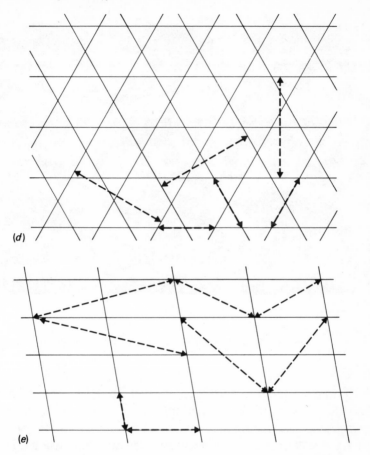

Fig. 3.8
2-dimensional lattices with several minimum displacement intervals indicated
for each by double-headed arrows

a 2-dimensional *lattice*. The patterns in fig. 3.8 are examples of lattices.
In each one the minimum displacement intervals for various directions
are indicated. Imaginative examples of 2-dimensional lattices have
been drawn by M.C. Escher. Many are collected in MacGillavry (ART,
6) and a few are included in Escher (ART, 2). A sample is presented
in fig. 3.9.

Conditions for approximate displacement symmetry in finite 2-di-
mensional systems are the same as in the 1-dimensional case. Common
examples of approximate planar displacement symmetry are: tiled
floors, graph paper, wallpaper patterns, window screen, planted forests.

Group theory. The set of all displacement transformations in a plane,

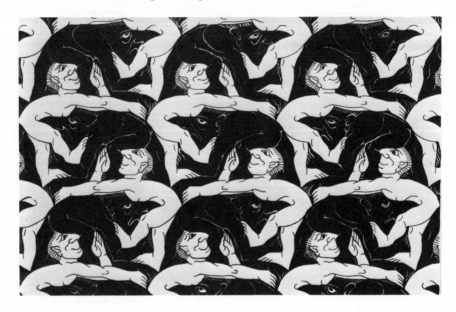

Fig. 3.9
2-dimensional lattice by M.C. Escher. (Collection Haags Gemeentemuseum, The Hague)

including the identity transformation, with the composition of consecutive displacement, form an infinite order commutative group, the *continuous planar displacement group*. Among these the set of symmetry displacement transformations of any 2-dimensional lattice forms an infinite order subgroup, a *discrete planar displacement group*. All such subgroups are isomorphic.

The plane reflection transformation is easily generalized to 2-dimensional systems by imagining a mirror with its plane perpendicular to the plane of the system and reflecting the system in it (fig. 3.10). The intersection of the mirror's plane with the system's plane is called the *reflection line*. Reflection transformations through different reflection lines do not commute.

An abstract plane has reflection symmetry through any reflection line lying in the plane. A circle has reflection symmetry through any reflection line lying in its plane and passing through its center (fig. 3.11). The forms in fig. 3.12 all have plane reflection symmetry. Their lines of reflection symmetry are indicated by dashed lines.

Examples of 2-dimensional systems with one or more lines of

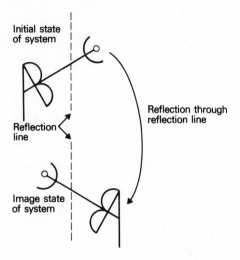

Fig. 3.10
Plane reflection transformation of 2-dimensional system

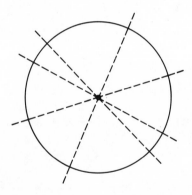

Fig. 3.11
Circle with several lines of reflection symmetry indicated by dashed lines

reflection symmetry are: blank or lined writing paper, Rorschach ink-blots, certain national flags, chess boards, many leaves.

PROBLEM

Where are the lines of reflection symmetry in each of these examples? Try to find additional examples.

Group theory. The identity transformation and reflection through

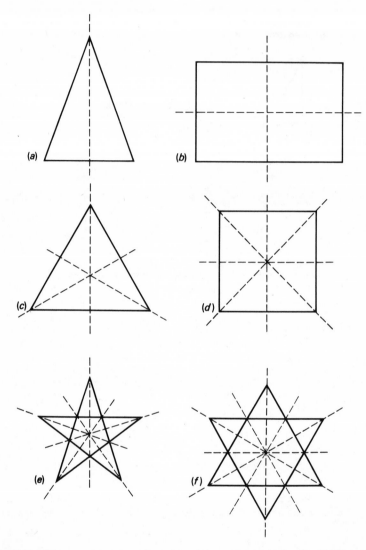

Fig. 3.12
Plane reflection symmetric 2-dimensional systems with lines of reflection symmetry indicated by dashed lines. (*a*) Isosceles triangle. (*b*) Rectangle. (*c*) Equilateral triangle. (*d*) Square. (*e*) Pentagram. (*f*) Hexagram ('Star of David')

a given reflection line, with consecutive reflection as composition, form a group of order 2.

Lattices can also have plane reflection symmetry. In analogy with the 1-dimensional case, if a lattice has a line of reflection symmetry, then this line must be one of an infinite set of parallel lines separated by half the minimum displacement interval for the lattice direction perpendicular to these lines. Examples of lattices with lines of reflection symmetry are fig. 3.8(*a*)–(*d*). The lines of reflection symmetry for each of these are exhibited in fig. 3.13. Only essentially different lines are indicated; all others are obtained from them by appropriate displacement. The lattice of fig. 3.8(*e*) possesses no plane reflection symmetry at all.

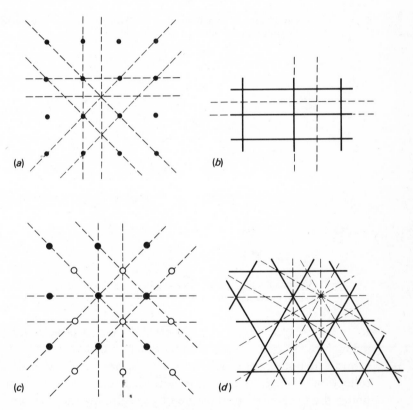

Fig. 3.13
Lines of reflection symmetry for those 2-dimensional lattices of fig. 3.8 that possess reflection symmetry, indicated by dashed lines. Only essentially different lines are shown

Consider fig. 3.14. It is symmetric under displacements in the direction of its length by interval *a* and multiples of *a*. It is also reflection symmetric through an infinity of reflection lines perpendicular

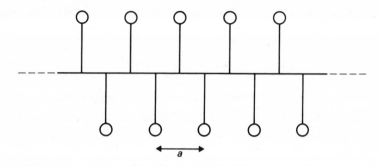

Fig. 3.14
Glide symmetric 2-dimensional system with single line of glide symmetry. Minimum displacement interval is indicated by *a*

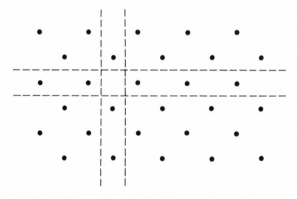

Fig. 3.15
Glide symmetric 2-dimensional lattice with several lines of glide symmetry indicated by dashed lines

to its length and separated by interval $\frac{1}{2}a$. (Each reflection line passes through the length of one of the 'branches' of the figure.) It is not symmetric under either longitudinal displacements by interval $\frac{1}{2}a$ or reflection through its central longitudinal axis. But it is symmetric under the transformation formed by combining these latter two transformations (in either order, since they commute). The whole symmetry transformation is then displacement along the axis by $\frac{1}{2}a$ accompanied by reflection through the axis. This transformation is called a *glide* transformation, and symmetry with respect to it is correspondingly called glide symmetry. The line involved in the transformation is called a *glide line*. The lattice of fig. 3.15 is an additional example of glide symmetry with some of the lines of glide symmetry indicated by dashed lines. The lattices of fig. 3.8(*c*), (*d*) also possess glide symmetry.

Find the lines of glide symmetry for these lattices.

A 2-dimensional system can undergo the transformation of *rotation*. This is achieved by rotating the whole system by a given angle about an axis perpendicular to its plane (fig. 3.16). The intersection of this axis with the plane is called the *rotation center* (or *rotocenter*). Rotations about the same center commute, but rotations about different centers do not.

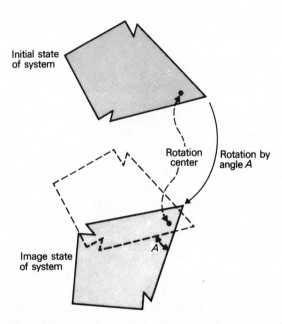

Initial state
of system

Rotation
center

Rotation by
angle *A*

A

Image state
of system

Fig. 3.16
Rotation transformation of 2-dimensional system

A circle is an example of a system that is symmetric under rotation by any angle about its center (fig. 3.17). Such a rotation center is called a center of *full rotational symmetry.*

Any system having a regular circlewise repetition is symmetric under rotation, but only by a certain minimum angle and multiples of it. This minimum rotation angle must be $360°/n$, where n is an integer greater than 1 (fig. 3.18). The corresponding rotation center is called

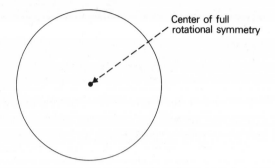

Center of full
rotational symmetry

Fig. 3.17
Circle is example of 2-dimensional system with full rotational symmetry

a center of *n-fold rotational symmetry*. (Note that by putting $n = \infty$ we obtain full rotational symmetry.)

PROBLEM

Why is n limited to being an integer?

All regular polygons have rotational symmetry about their centers. The equilateral triangle and the square (fig. 3.18(b), (c)) have respectively 3-fold and 4-fold rotational symmetry. A regular pentagon or pentagram (fig. 3.18(d)) has a center of 5-fold rotational symmetry. And a general regular n-sided polygon (called a regular n-gon in mathematical shorthand) or n-pointed star (n-gram) has n-fold rotational symmetry about its center. Refer to Coxeter (GEO, 1) and Steinhaus (POLY, 3).

Group theory. The set of transformations consisting of the identity transformation and rotations about a given rotation center by $k \times 360°/n$, with n a given positive integer and $k = 1, 2, \ldots, n-1$, and with consecutive rotation as composition, forms a commutative group called the *cyclic group* of order n and denoted C_n. Thus C_1 is just the identity transformation, and C_2, C_3, C_5 are realizations of the abstract groups of orders 2, 3, 5, respectively, as presented in chapter 2. C_4 is the realization of the first group of order 4 there. For $n \geqslant 3$, C_n is the rotational (but not the full) symmetry group of the regular n-gon and n-gram.

2-dimensional systems might be simultaneously symmetric under both rotations and reflections. Then a center of n-fold rotational symmetry will have n lines of reflection symmetry passing through it, mutually separated by half the minimum rotation angle.

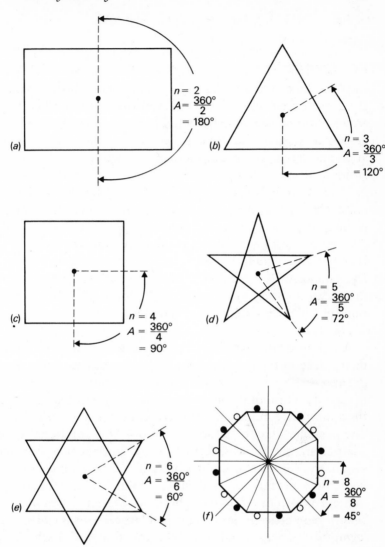

Fig. 3.18
2-dimensional systems with *n*-fold rotational symmetry. Minimum rotation angle
is $A = 360°/n$. (*a*) Rectangle. (*b*) Equilateral triangle. (*c*) Square. (*d*) Pentagram.
(*e*) Hexagram ('Star of David'). (*f*) Form based on regular octagon

PROBLEM

Why is this so?

All regular *n*-gons and *n*-grams possess such symmetry. The lines of
reflection symmetry for the examples of fig. 3.18(*a*)–(*e*) are shown

in fig. 3.12(*b*)–(*f*). Figure 3.18(*f*) has 8-fold rotational symmetry but no reflection symmetry.

Group theory. The set consisting of all the transformations of C_n together with reflections through *n* reflection lines passing through the rotation center and mutually separated by $180°/n$, with consecutive transformation as composition, forms a group called the *dihedral group* of order 2*n* and denoted D_n. Except for $n = 1$ and 2, these groups are noncommutative. D_1 consists of the identity transformation and reflection through a single reflection line. It is the symmetry group of Rorschach inkblots. D_2 is the realization of the second group or order 4 presented in chapter 2 and is the symmetry group of the rectangle. D_3 is a realization of the group of order 6 shown there and is the symmetry group of the square. (There is one other order 6 group, realized by C_6.) D_n for $n \geqslant 3$ is the full symmetry group of the regular *n*-gon and *n*-gram. Figure 3.18(*f*) possesses C_8 symmetry but not D_8 symmetry. The groups C_n and D_n for all *n* make up what are called the *planar point groups*. These are all groups of rotations and reflections in a plane leaving a point fixed. Refer, for example, to Weyl (SYM, 1) or Coxeter (GEO, 1).

A 2-dimensional lattice can also be endowed with rotational symmetry. Due to the displacement symmetry of the lattice, every center of rotational symmetry will be repeated an infinite number of times in each direction of displacement symmetry at a separation equal to the appropriate minimum displacement interval for each such direction. The possible values of *n* for a center of *n*-fold rotational symmetry in a lattice are severely restricted by the requirement of displacement symmetry. Geometric compatibility of the two kinds of symmetry limits these centers to being either 2-fold, 3-fold, 4-fold or 6-fold. Centers of 5-fold, 7-fold or higher-fold rotational symmetry are simply geometrically impossible in a lattice. This is called the *crystallographic restriction* and is well illustrated in M.C. Escher's periodic drawings (ART, 2, 6).

PROBLEM

You might attempt to prove the crystallographic restriction. Weyl (SYM, 1), Coxeter (GEO, 1), Jaswon (CRYS, 3) and Nussbaum (PHYS, 12), among others, give proofs.

All the examples of 2-dimensional lattices that we presented in fig. 3.8 happen to have rotational symmetry. In fig. 3.19 we exhibit the

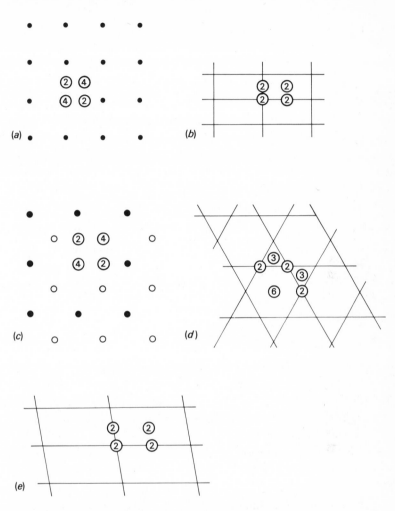

Fig. 3.19
Centers of rotational symmetry for 2-dimensional lattices of fig. 3.8. (n) denotes
n-fold center. Only essentially different centers are shown

centers of rotational symmetry for each of these lattices. We indicate
only the essentially different centers; all others are obtained from
them by appropriate displacement. An n-fold center is denoted by (n).
Figure 3.20 is an example of a 2-dimensional lattice with no rotational
symmetry at all.

As we mentioned, the displacement symmetry of a 2-dimensional
lattice imposes restrictions on the reflection and rotational symmetries
that the lattice may possess. And, it is readily guessed, the coexistence

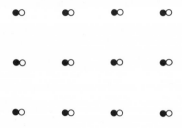

Fig. 3.20
2-dimensional lattice with no rotational symmetry at all

of reflection and rotation symmetries imposes additional restrictions on both. Thus are the possible lattice symmetries involving displacement (of course), reflection (and glide) and rotation transformations severely limited. They are so limited, in fact, that the different symmetries of this kind, called *planar crystallographic space groups*, can be fairly easily determined and counted. There are seventeen. See Weyl (SYM, 1), Coxeter (GEO, 1), Jaswon (CRYS, 3), Wells (CRYS, 5) and Loeb (COL, 2).

Group theory. The planar crystallographic space groups are indeed groups in the mathematical sense. They are the only transformation groups possible as full symmetry groups of 2-dimensional lattices. The *planar crystallographic point groups* are those planar point groups that are subgroups of planar crystallographic space groups; that is, they are those planar point groups that can be symmetries of a 2-dimensional lattice. Which are they?

The transformation of *line reflection* (or *line inversion*) acts on a 2-dimensional system by reflecting each point of the system through a line perpendicular to the plane of the system. The intersection of the line and the plane is called the *reflection center*. The image of a point is therefore located in the plane of the system, directly opposite the initial point with respect to the reflection center, and at the same distance from the reflection center as is the initial point (fig. 3.21). Reflection transformations through different reflection centers do not commute.

We do not devote further discussion to this transformation, because it is not really new; a line reflection transformation is completely equivalent to rotation by 180° about the reflection center.

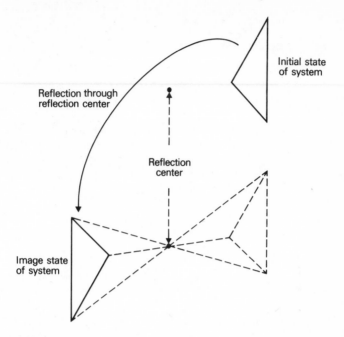

Fig. 3.21
Line reflection transformation of 2-dimensional system

PROBLEM

Convince yourself of this fact.

Spatial symmetry

Pooh looked at his two paws. He knew that one of them was
the right, and he knew that when you had decided which one
of them was the right, then the other one was the left, but he
never could remember how to begin.
(A.A. Milne: *The House at Pooh Corner*)

All the geometric transformations that we considered for 2-dimensional
systems are readily applied to 3-dimensional systems. These transformations are displacement, plane and line reflection, glide and rotation.
As expected, the addition of a third dimension allows the application
of transformations that are inapplicable to lower-dimensional systems.
Of these we discuss point reflection and the screw. We also consider
the dilation transformation, applicable to systems of any number of
dimensions.

The action of the displacement transformation on 3-dimensional
systems is the obvious generalization of its action on 2-dimensional

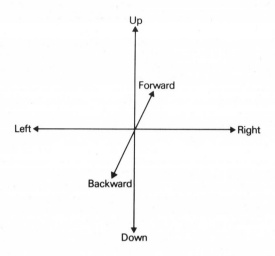

Fig. 3.22
Some possible displacement transformations for 3-dimensional systems. 'Back-ward–forward' line is perpendicular to plane of page

systems; there is one more dimension to displace into (fig. 3.22). As in the 2-dimensional case, all displacement transformations commute, whether or not they are in the same direction.

The conditions for exact displacement symmetry of 3-dimensional systems are the same, with appropriate generalization, as in the 2-dimensional case.

An abstract unbounded 3-dimensional space (for those readers familiar with the concept of space curvature, we must qualify this space to be flat) has displacement invariance in any direction by any interval.

A 3-dimensional system that has infinite regular repetition in at least three independent directions (that is, three directions that are not all parallel to the same plane) is called a 3-dimensional lattice. This is an obvious generalization of a 2-dimensional lattice. In fact, a 3-dimensional lattice can be viewed as an infinite regular pile of identical sets of 2-dimensional lattices. (More than one kind of 2-dimensional lattice may contribute to the structure of a single 3-dimensional lattice.) Since it is difficult to draw 3-dimensional lattices clearly, we use our imagination in constructing examples and picture infinite regular piles of each of the examples of 2-dimensional lattices in fig. 3.8. It is important to note that pairs of corresponding points in consecutive identical layers do not have to lie directly above and below each other but may be relatively displaced in any direction parallel to the layers. This

Fig. 3.23
Simple 3-dimensional lattice

relative displacement must then be repeated from layer to layer.
Figure 3.23 shows an example of a simple 3-dimensional lattice as a
pile of identical 2-dimensional lattices with a relative displacement of
neighboring layers.

Approximate displacement symmetry in finite 3-dimensional systems occurs under the same conditions as were described for 2- and
1-dimensional systems. The most important and common examples
of approximate spatial displacement symmetry are crystals. These are
3-dimensional lattices built out of atoms, ions or molecules. Since
crystals are finite systems, they are actually not lattices but only parts
of lattices (which are infinite by definition), but there is no harm in
using the term a little loosely. Refer to Holden and Singer (CRYS, 2),
Mott (CRYS, 4), Wells (CRYS, 5) and Holden (CRYS, 6).

PROBLEM

Find additional examples of finite 3-dimensional lattices.

Group theory. The set of all displacement transformations in space,
including the identity transformation, with consecutive displacement
as composition, forms an infinite order commutative group, the *continuous spatial displacement group.* Among these the set of symmetry
displacement transformations of any 3-dimensional lattice forms an
infinite order subgroup, a *discrete spatial displacement group.* All such
subgroups are isomorphic.

The rotation transformation acts on a 3-dimensional system by rotating the whole system through a given angle about a given axis, called the *rotation axis*. Rotations about the same axis commute, but rotations about different axes do not.

As in the 2-dimensional case, a 3-dimensional system might be symmetric under rotations by any angle about one or more axes or only by a minimum angle of 360°/n, where n is an integer greater than 1, and multiples of it. Such a rotation axis is then called an axis of *full* or *n-fold rotational symmetry*, respectively.

A sphere, for example, has an infinite number of axes of full rotational symmetry — all axes passing through its center (fig. 3.24). Systems having this type of symmetry are said to possess *spherical symmetry*. A cylinder has full rotational symmetry only about a single axis, the longitudinal axis of the cylinder. Also a cone possesses this symmetry. This is referred to as *axial symmetry* (fig. 3.25).

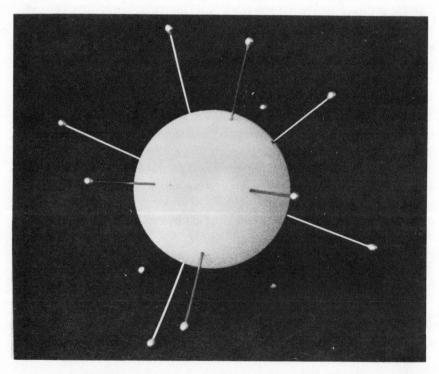

Fig. 3.24
Sphere is example of 3-dimensional system with spherical symmetry. A number of axes of full rotational symmetry are indicated

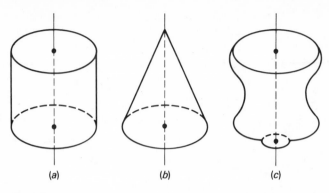

Fig. 3.25
3-dimensional systems with axial symmetry

An infinitely long cylinder with no ends has, in addition to its axial symmetry, displacement symmetry by any interval in the direction of its axis. This combination of symmetries, displacement symmetry by any interval along an axis of axial symmetry, is called *cylindrical symmetry*.

Examples of spherical symmetry are: balls, ball bearings, Edam cheeses. Examples of axial symmetry: round jars, round undecorated (unless the decoration is appropriately symmetric) dinner plates, bullets, any object turned on a lathe. Finite systems cannot possess cylindrical symmetry, so we cannot offer concrete examples in this case.

PROBLEM

Think of more examples.

Group theory. Rotations about a common axis give us again the cyclic groups C_n. For $n \geqslant 3$, C_n is the rotational (but not full) symmetry group of the pyramid built on a regular n-gonal base (with axis of rotational symmetry passing through the apex and the center of the base).

A regular tetrahedron has three axes of 2-fold rotational symmetry (through the midpoints of pairs of opposite edges) and four 3-fold rotational symmetry axes (through each vertex and the center of the opposite face) (fig. 3.26(*a*)). A cube has six 2-fold axes (through the midpoints of pairs of opposite edges), four 3-fold axes (through pairs of opposite vertices), and three 4-fold axes (through the centers of pairs of opposite faces) (fig. 3.26(*b*)).

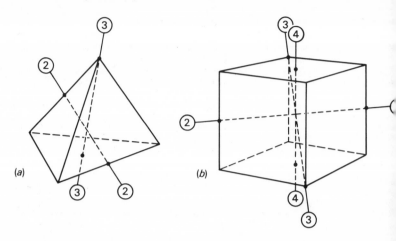

Fig. 3.26
Representative rotational symmetry axes of (*a*) regular tetrahedron and (*b*) cube.
ⓝ indicates *n*-fold axis

PROBLEM

The regular tetrahedron and the cube are two of a series of
five polyhedrons called Platonic solids. The other three are
the regular octahedron, the regular dodecahedron and the
regular icosahedron (fig. 3.27). The regular tetrahedron,
octahedron and icosahedron are formed of four, eight and
twenty (respectively) identical equilateral triangles, with
three, four and five (respectively) triangular faces meeting
at each vertex. The cube is formed of six identical squares
with three faces meeting at each vertex. The regular dodecahedron is formed of twelve identical regular pentagons with
three faces meeting at each vertex. The characteristics of the
Platonic series are that for each member (*a*) all faces are identical regular polygons, (*b*) all vertices are identical, and (*c*) each
member is everywhere convex. Why does this series have no
more members? Hint: For each regular polygon consider
how many identical ones can meet at a vertex. Carry through
the rotational symmetry analysis, that we performed for the
regular tetrahedron and the cube, for the rest of the series.
You might construct the whole series and indicate all rotational symmetry axes and their types directly on your
models. Compare the rotational symmetry of the cube with
that of the octahedron, and that of the dodecahedron with

Fig. 3.27

(*a*) Regular octahedron

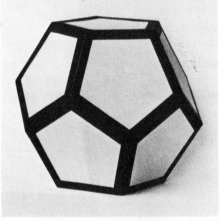

(*b*) Regular dodecahedron

(*c*) Regular icosahedron

that of the icosahedron. Explain. How does the tetrahedron differ from these four?

References, including instructions for construction of these and other polyhedrons, are listed under POLY in the bibliography, together with (GEO, 1) and (CRYS, 5).

As in the 2-dimensional case, a 3-dimensional lattice can possess rotational symmetry with respect to one or more sets of parallel axes, but with the additional freedom that the sets may have various orientations. The considerations of the 2-dimensional case, including the crystallographic restriction, the limitation to 2-fold, 3-fold, 4-fold or 6-fold rotational symmetry, apply here as well.

The transformations of plane and line reflection act on 3-dimensional systems in a manner that is a suitable generalization of their action on 2-dimensional systems.

For the plane reflection transformation a plane, the *reflection plane*, must first be chosen. Then the system is reflected in this plane as if the plane were a two-sided mirror (fig. 3.28). In geometric terms, we find the image of any point of the system by dropping a perpendicular from the point to the reflection plane and continuing the line the same distance on the opposite side of the plane. The image is then located at the other end of this line segment.

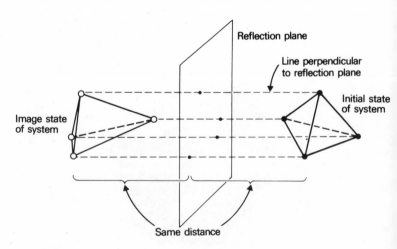

Fig. 3.28
Plane reflection transformation of 3-dimensional system

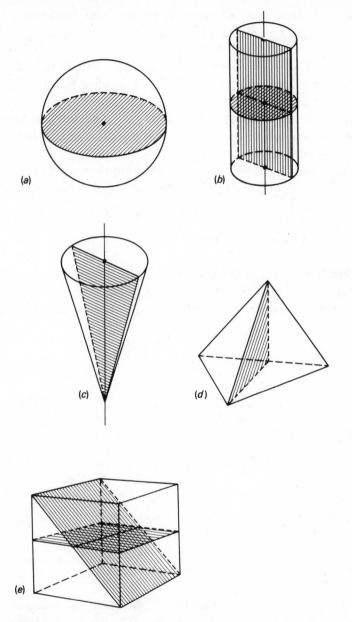

Fig. 3.29
3-dimensional systems with plane reflection symmetry. Representative planes of reflection symmetry are shown. (*a*) Sphere. (*b*) Cylinder. (*c*) Cone. (*d*) Regular tetrahedron. (*e*) Cube

A sphere has plane reflection symmetry with respect to any plane through its center (fig. 3.29(*a*)). A cylinder or cone has plane reflection symmetry with respect to any plane through its axis (fig. 3.29(*b*), (*c*)). A cylinder, but not a cone, has an additional plane of reflection symmetry perpendicular to its axis and dividing it into two equal parts. A regular tetrahedron has six planes of reflection symmetry (all of the same kind) (fig. 3.29(*d*)), and a cube has six of one kind (passing through two opposite edges) and three of another (parallel to a pair of opposite faces) (fig. 3.29(*e*)).

Additional examples of plane reflection symmetry are the general external appearance of: cars, many public buildings, most mammals.

PROBLEM

Find more examples.

The line reflection transformation is performed by choosing a line, the *reflection line*, and reflecting the system through this line (fig. 3.30). We find the image of each point of the system by dropping a

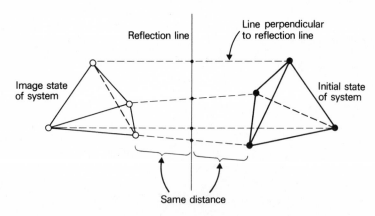

Fig. 3.30
Line reflection transformation of 3-dimensional system

perpendicular from the point to the reflection line and continuing it on past the line for the same distance. The end of this line segment then marks the location of the image point. As in the 2-dimensional case, the line reflection transformation is completely equivalent to rotation by 180° about the reflection line.

The transformation of *point reflection* (or *point inversion*) is an

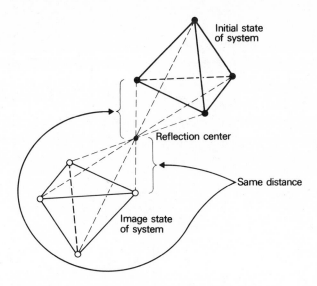

Fig. 3.31
Point reflection transformation of 3-dimensional system

essentially 3-dimensional transformation (fig. 3.31). To perform a point reflection we pick a point, the *reflection center*, and reflect the system through this point by running a line from each point of the system to the reflection center and on past it for the same distance. The image point is then located at the far end of this line segment.

A sphere, a cylinder and a cube, for example, possess point reflection symmetry with respect to their centers (fig. 3.32(*a*)-(*c*)). Tetrahedrons and cones are not point reflection symmetric. But a double cone, consisting of two identical cones placed base to base or apex to apex with their axes coinciding, has a center of reflection symmetry (fig. 3.32(*d*), (*e*)).

Group theory. Any reflection transformation (whether through a plane, a line or a point) and the identity transformation, with consecutive reflection as composition, form an order 2 group.

Let us imagine now a plane, a line perpendicular to it, and their point of intersection (fig. 3.33). Consider the transformations of plane reflection in the plane, line reflection through the line (which is the same as rotation by 180° about the line) and point reflection through the point of intersection. First of all, these three transformations commute with each other.

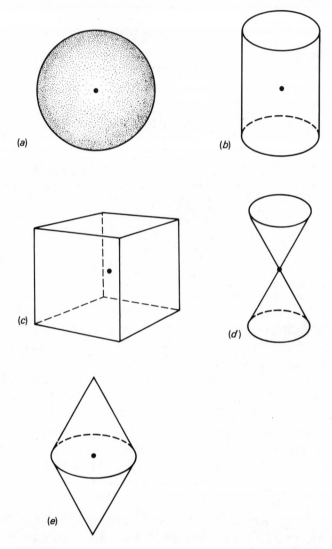

Fig. 3.32
3-dimensional systems with point reflection symmetry. Center of reflection symmetry is indicated for each by dot. (*a*) Sphere. (*b*) Cylinder. (*c*) Cube. (*d*) Double cone. (*e*) Double cone

PROBLEM

Show that this is true.

What we want to point out mainly, however, is that the result of

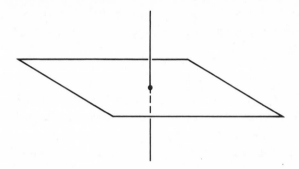

Fig. 3.33
Perpendicular line and plane and their point of intersection

consecutively applying any two of these three transformations to a system is the same result as would have been obtained by applying the third transformation instead. For example, the plane reflection transformation followed by the line reflection (or vice versa) give the point reflection transformation.

PROBLEM

Prove generally or by examining each of the three cases that these transformations are in fact interrelated as described.

It follows, of course, that if a system is symmetric under any two of these reflection transformations, it must also be symmetric under the third. Examples of systems possessing this symmetry are spheres, cylinders, cubes and both kinds of double cones (fig. 3.32).

PROBLEM

Find all pairs of plane and perpendicular axis of reflection symmetry for these examples.

Group theory. These three reflection transformations together with the identity transformation, with consecutive transformation as composition, form another realization of the second group of order 4 in chapter 2.

PROBLEM

Find the reflection symmetries of all the Platonic solids.

Group theory. The *spatial point groups* are all the groups of rotations and reflections in space that leave a point fixed. We do not enumerate them or otherwise go into details, but just mention that they include, among others, the cyclic groups C_n, spatial generalizations of the dihedral groups D_n, and the symmetry groups of the Platonic solids. See Coxeter (GEO, 1) or Weyl (SYM, 1).

The *glide* transformation has a straightforward generalization to 3-dimensional systems. The system is displaced parallel to a *glide plane* and reflected in this plane.

The *screw* transformation is presented through an example. Consider a spiral (more properly called a helical) staircase. Let h denote the change of height involved in each complete turn of the staircase and n the number of steps in one complete turn. The height of each step above the one below it is then h/n (fig. 3.34). The spiral staircase has no rotational symmetry about is axis. If it were infinitely long, it

Fig. 3.34
Spiral staircase, side view

would have displacement symmetry along its axis with minimum displacement interval h.

Since there are n steps per turn, each step is essentially a wedge of angle $360°/n$ (fig. 3.35). So a rotation of the staircase about its axis

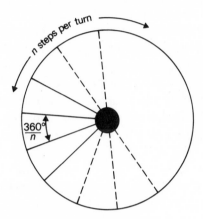

Fig. 3.35
Spiral staircase, view along axis

by $360°/n$ puts each step exactly in a position either above where the step below it was or below where the step above it was, depending on which way the rotation is made. Now, obviously, a displacement by interval h/n along its axis can return the (infinite) staircase to its original appearance. The combined transformation then leaves each step just where the next (higher or lower) one was, and an infinite staircase is invariant under this transformation, which is an example of a screw transformation.

The general definition of a screw transformation is a transformation consisting of consecutive application (in either order) of a rotation and a displacement along the axis of rotation, now called a *screw axis*. Screw transformations along the same screw axis commute.

Systems possessing screw symmetry must be of general helical form and of infinite length. They might be step-like or continuous. An infinite spiral staircase is an example of the former type, and an infinite screw-thread (the origin of this transformation's name) of the latter. In the step-like case the system is invariant only under a certain minimum screw transformation, consisting of a minimum rotation by angle $360°/n$ and minimum displacement by interval h/n, and multiples of this minimum screw transformation. In the continuous case the screw transformations under which the system is symmetric may consist of a rotation by any angle $A°$ accompanied by a displacement through interval $h \times A°/360°$.

PROBLEM

Prove this for the continuous case.

In addition to rotational symmetry, 3-dimensional lattices can possess plane reflection, point reflection, glide and screw symmetry. As in the 2-dimensional case the coexistence of these symmetries with the lattice's displacement symmetry and with each other imposes restrictions on them all. And again the possible lattice symmetries involving displacement (by definition), reflection (and glide) and rotation (and screw) transformations are limited. The different symmetries of this kind, called *spatial crystallographic space groups*, can be determined with quite a bit of effort. They number 230. These symmetries are divided into 32 classes, called *crystallographic point groups*. The classes are further divided into seven (sometimes into six) crystallographic systems. Refer to Weyl (SYM, 1), Coxeter (GEO, 1), Dorain (PHYS, 1), Jaswon (CRYS, 3), Wells (CRYS, 5) and Shubnikov (COL, 1).

The *dilation* transformation is a simple one. It involves only changing all lengths of a system by the same factor (fig. 3.36). Doubling all lengths of a system or halving them are two examples of dilation

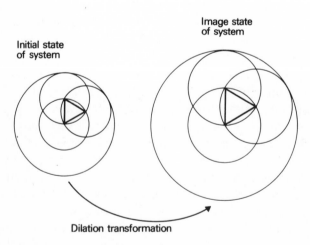

Image state
of system

Initial state
of system

Dilation transformation

Fig. 3.36
Dilation transformation

transformations. This transformation is sometimes called a *scale* transformation and is clearly applicable to systems of one, two and three
dimensions. Dilation transformations commute with each other. Shapes
are not affected by dilation transformations. A sphere becomes a
larger or smaller sphere; a cube is transformed into a larger or smaller
cube; etc.

Only infinite systems can have dilation symmetry. An infinite straight
line is symmetric with respect to all dilation transformations. An infinite set of circles, all having a common center and lying in the same
plane, where the radius of each circle is twice the radius of the next
smaller circle and half that of the next larger, is symmetric with respect to dilation transformations multiplying or dividing lengths by
powers of 2; that is, by 2, 4, 8, 16, etc. (fig. 3.37).

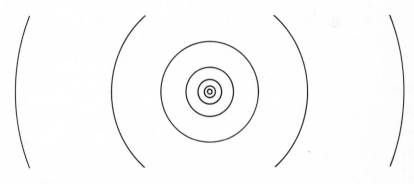

Fig. 3.37
Part of system of circles that is symmetric under dilation by powers of 2

PROBLEM
Show that this is so.

To conclude this chapter we briefly remark on interconnections
between displacement transformations, on the one hand, and cosmology
and the study of the extremely small, on the other. By how large an
interval is it possible to perform a displacement? This is actually a cosmological question, and the answer to it is intimately linked to the
structure of the universe. We cannot enter into a detailed discussion
here. However, we mention that, if the universe is spatially 'open',
there is no limit to a displacement interval. If, on the other hand, the
universe is spatially 'closed', nature imposes a natural limit on

displacement intervals. A 'closed' universe can be compared to the surface of a sphere, though it must be kept in mind that the former is 3-dimensional while the spherical surface is 2-dimensional. A spaceship traveling in what its navigator believes to be a straight line within a 'closed' universe, like an ant walking in its 'straight line' (a great circle) on a sphere's surface, eventually returns to its point of departure. Thus the 'circumference' of the 'closed' universe is the limit that nature imposes on displacements. According to this picture displacements are not really displacements, but have a rotational character. The question of whether the universe is in fact spatially 'open' or 'closed' awaits the verdict of the astronomers. References on cosmology are listed in the bibliography under COS, including some of the cross-references.

It is equally valid to inquire by how small an interval a displacement can be performed. In other words, is there in nature a fundamental minimum length, such that smaller lengths are in some sense meaningless? This is a completely open question at present and has bearing on current research in the physics of elementary particles.

4
Other symmetries and approximate symmetry

Temporal symmetry

He looked up at his clock, which had stopped at five minutes
to eleven some weeks ago.
'Nearly eleven o'clock,' said Pooh happily. 'You're just in
time for a little smackerel of something'. . .
The clock was still saying five minutes to eleven when
Pooh and Piglet set out on their way half an hour later.
(A.A. Milne: *The House at Pooh Corner*)

After looking into geometric symmetry with its wealth of symmetry
transformations, we turn to a consideration of *temporal symmetry*.
Here we are concerned with symmetry under transformations involving time. Whereas geometric symmetry has to do with the geometric
properties of systems at any given instant, temporal symmetry refers
to the development of systems as time progresses. For general references on time refer to Schlegel (TIM, 6) and Goudsmit, Claiborne
and Time–Life editors (TIM, 3).

Time can be likened to a line. Just as the specification of a single
number (coordinate) is sufficient to designate the location of a point
on a line, so is the occurence of·a 'point in time', an instant, fixed by
a single number (time coordinate), its 'time' or 'o'clock', which might
be expressed in terms of years, weeks, hours, seconds, etc. So we
should be able to apply our discussion of linear geometric symmetry
in the previous chapter to temporal symmetry, which is what we in
fact do. The temporal transformations that we consider are displacement, reflection and dilation.

The transformation of *temporal displacement* (or *time displacement*
or *translation*) involves displacing the system in time by a given time
interval (fig. 4.1). For example, if the system is a person's daily routine,
and he moves to another time zone and adjusts his schedule accordingly,
he performs a temporal displacement on his routine by a number of
hours equaling the difference between time zones. He must also temporally displace his clocks by the same amount. Or if one dozes an

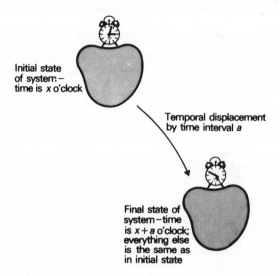

Initial state
of system: —
time is *x* o'clock

Temporal displacement
by time interval *a*

Final state of
system–time
is *x* + *a* o'clock;
everything else
is the same as
in initial state

Fig. 4.1
Temporal displacement transformation

extra ten minutes after the alarm clock sounds, he displaces his rising
by ten minutes. In analogy with the geometric case, a system can be
symmetric under temporal displacement only if it exists for an infinite
duration and has neither beginning nor end; that is, if it is eternal.

An eternal static (or temporally homogeneous) system is symmetric
under temporal displacements by any time interval. Since it is static,
it does not change, and there is nothing to distinguish its state at any
time from its state at any other time. An eternally existing system
whose behavior is periodic, repeating itself after a constant time interval called its *period*, is symmetric under temporal displacements, but
only under displacements by time intervals equaling its period or
multiples of it. The period of temporal symmetry is analogous to the
minimum displacement interval of geometric symmetry.

Since no real systems are eternal (we put cosmological considerations
aside for the present), no real systems have exact temporal displacement symmetry. But, again in analogy with geometric symmetry, a
system of finite duration (which we might call a mortal system) is
said to have approximate temporal displacement symmetry, if its
duration can be imagined as being part of the infinite duration of an
exactly symmetric system, and if the period is much less than the total
duration of the system. Then we can ignore the effect of the beginning and end, on condition that we observe the system at sufficiently

long times after its beginning and before its end and restrict the temporal displacements to sufficiently small multiples of the period.

Examples of systems with approximate temporal displacement symmetry are many. A clock is such a system, whether its period is taken to be the period of oscillation of its balance wheel ($\frac{2}{5}$ of a second in most watches), the period of rotation of its hour hand (twelve hours), or the time interval between windings (usually 24 hours). Among astronomical phenomena we have the revolution of planets about their suns with a period of about $365\frac{1}{4}$ days for the earth, the revolution of satellites about their planets with a sidereal period of about $27\frac{1}{3}$ days for the earth's moon, the rotation of planets about their axes with a sidereal period of about 23 hours and 56 minutes for the earth, pulsars (pulsating stars) with periods of a fraction of a second (COS, 11), and many more. A steady tone is a system with approximate temporal displacement symmetry. Its period is the period of the vibrations producing the tone. Another example is a record player turntable in operation. Its period is usually $1/33\frac{1}{3}$ of a minute, but might be 1/45 or 1/78 of a minute, depending on the kind of record being played.

PROBLEM

Find additional examples.

The *temporal reflection* (or *time reflection, inversion* or *reversal*) transformation is accomplished by having the system run in reverse; that is, we reverse the order of the events making up the development of the system in such a way that the time intervals between successive events retain their magnitudes. The usual means of illustrating this transformation is to project a motion picture with the film running backward (fig. 4.2). A rotating wheel will turn the other way, a falling body will fly upward, people will grow younger, etc. Sounds can be temporally reflected by running a recording of them backward through a tape recorder. 'We' becomes 'you', 'dog' becomes 'god', 'say' becomes 'yes', and vice versa.

PROBLEM

Experiment with temporal reflection using a motion picture projector that can run in reverse and a tape recorder on which tapes can be played in reverse. This cannot usually be done on ordinary two-track tape recorders.

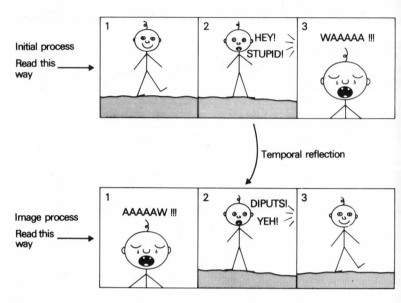

Fig. 4.2
Example of temporal reflection transformation

Systems possessing temporal reflection symmetry must have in their development one instant, for systems of finite duration, or at least one instant, for eternal systems, when the system starts to retrace in reverse order all the previous steps of its development. This *turning point* is analogous to the center of reflection symmetry in 1-dimensional systems. Conversely, if a system has a turning point, then it is symmetric under temporal reflection.

If an eternal system has more than one turning point, it has an infinity of turning points, separated by equal time intervals. The behavior of the system during any such interval must be the temporal reflection of its behavior during the preceding or following interval. So such a system is periodic with a period equal to twice the interval between successive turning points. Then, as we saw above, it has temporal displacement symmetry. Conversely, if a temporal displacement symmetric system also has temporal reflection symmetry, it has an infinite number of turning points separated by half the period.

PROBLEM

Convince yourself of the correctness of every statement made in the preceding paragraph.

An example of a system possessing temporal reflection symmetry is the sound of the sentence, 'Say yes.' Another is the motion of a perfectly elastic ball dropped from rest on to a level surface, from the instant of release to the instant of its return to rest at its initial position, after any number of bounces. This example becomes one of approximate temporal displacement symmetry, if the ball is allowed to bounce many times.

PROBLEM

Look for additional examples.

The transformation of *temporal dilation* (or *time dilation*) changes all time intervals in the development of a system by the same factor. Doubling all time intervals or halving them are two examples of temporal dilation transformations. In music these transformations are called augmentation and diminution, respectively, and are two of many ways of modifying a melodic theme (fig. 4.3). Refer to Copland (TIM, 8) and Mann (TIM, 9).

Fig. 4.3
Example of temporal dilation

Only eternal or semi-eternal (this is, having either no beginning or no end) systems can possess temporal dilation symmetry. An eternal static system is symmetric with respect to all temporal dilation transformations. An infinite set of identical events occurring at times . . . , $\frac{1}{64}, \frac{1}{32}, \frac{1}{16}, \frac{1}{8}, \frac{1}{4}, \frac{1}{2}, 1, 2, 4, 8, 16, 32, 64, \ldots$ (this set has a beginning but no end) is symmetric with respect to dilation transformations multiplying or dividing time intervals by powers of 2; that is, by 2, 4, 8, 16, etc.

PROBLEM

Show that the last example has the symmetry ascribed to it.

Now for some cosmological considerations. We ask by how large a time interval it is possible to perform a temporal displacement. And the answer to this question, just like the answer to the analogous question at the end of the previous chapter, ties in with cosmology. Even philosophers are concerned with this, because the question has bearing on the very nature of time: how 'long' is time? Most scientists who deal with these matters feel that the concept of time has meaning only relative to the existence of the universe, so we ask how long a time the universe has existed and how much time remains. Present evidence seems to indicate that the universe that we observe started as extremely dense and hot matter that exploded and has been expanding ever since. Time started at the instant of the 'big bang'. The eventual fate of the universe could probably be predicted, if only enough astronomical data were available, but at present this is a matter for speculation. However, two possibilities present themselves. The universe might continue expanding for ever, in which case temporal displacements into the future would be unlimited while those into the past would not. Or it might reach a state of maximum expansion and then retrace its development and contract back into a 'fireball', at which stage time would end. In this case all temporal displacements, whether into the future or into the past, would be limited. General references on cosmology are listed in the bibliography under COS, including the cross-references. Bonnor (COS, 19) is especially relevant.

It has even been suggested that time might reach its 'maximum' at the maximum expansion stage of the universe and would then 'run backward' during the contraction process until it returns to its beginning at the 'fireball' stage. The end would be the beginning and the beginning would be the end! There can be no question about what would happen 'before' the beginning or 'after' the end of time. Time would have a 'closed' structure, and temporal displacements would have the character of rotations. Compare this with the spatially 'closed' universe described at the end of the previous chapter. In such a universe no question about what lies 'outside' can arise, because there is no 'outside'. This is discussed by Gardner (TIM, 2).

And what about a lower limit to temporal displacements? Does nature impose a fundamental minimum time interval such that smaller time intervals are in some sense meaningless? This question is closely linked with the analogous question for spatial intervals and has no known answer at present.

Permutation symmetry

> . . . suddenly he remembered how he and Piglet had once
> made a Pooh Trap for Heffalumps, and he guessed what had
> happened. He and Piglet had fallen into a Heffalump Trap for
> Poohs!
>
> (A.A. Milne: *The House at Pooh Corner*)

The transformation of *permutation* is a rearranging or reordering of parts of a system. Symmetry with respect to this transformation is *permutation symmetry*. Refer to Weyl (SYM, 1) and Coxeter (GEO, 1).

Consider this example. Three depressions in the sand, labeled A, B, C, are occupied by three balls. Permutations of this system are specified by tables such as the following.

	Ball initially in depression	Is to be placed in depression
Permutation 1	$\begin{cases} A \\ B \\ C \end{cases}$	$\begin{matrix} C \\ B \\ A \end{matrix}$
Permutation 2	$\begin{cases} A \\ B \\ C \end{cases}$	$\begin{matrix} B \\ C \\ A \end{matrix}$

.
.
.

etc.

If all three balls are different, any permutation will change the system, so it has no permutation symmetry. If all three balls are similar, the system is clearly invariant with respect to any permutation, and therefore possesses full permutation symmetry. An intermediate case is also possible, when two of the balls are similar, while the third differs from them. The system still has permutation symmetry, but not under all possible permutations. For example, if the similar balls are initially in depressions A and C, then permutation 1 is a symmetry transformation, while permutation 2 is not.

Or consider an arbitrary triangle with sides of lengths a, b, c and angles A, B, C respectively opposite them as in fig. 4.4. We have the well-known sine rule,

$$\frac{a}{\sin A} = \frac{b}{\sin B} = \frac{c}{\sin C},$$

and three cosine rules,

$$a^2 = b^2 + c^2 - 2bc \cos A,$$

$$b^2 = c^2 + a^2 - 2ca \cos B,$$

$$c^2 = a^2 + b^2 - 2ab \cos C.$$

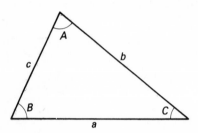

Fig. 4.4
Arbitrary triangle

The sine rule is symmetric under all permutations of the three pairs (a, A), (b, B), (c, C). Each cosine rule is symmetric only under the interchange of two of these pairs: the first rule only under the interchange of (b, B) and (c, C), the second only of (c, C) and (a, A), and the third only of (a, A) and (b, B). Each pair of cosine rules has the same permutation symmetry as the one not included in the pair, while the whole set of cosine rules possesses the full permutation symmetry of the sine rule.

In similar vein we have the formulae for the area of the triangle,

$$\text{area} = \sqrt{\{s(s-a)(s-b)(s-c)\}},$$

the radius of the inscribed circle,

$$r_{\text{insc}} = \frac{\text{area}}{s},$$

and the radius of the circumscribed circle (fig. 4.5),

$$r_{\text{circ}} = \frac{abc}{4 \times \text{area}},$$

where $s = \frac{1}{2}(a + b + c)$.

Each of these formulae is symmetric under all permutations of the three sides a, b, c.

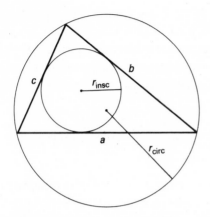

Fig. 4.5
Arbitrary triangle with inscribed and circumscribed circles

Permutation symmetry can also be more abstract. For example, among the possible relations between a pair of numbers, consider the relation of 'equality' (=). Two numbers, m and n, might be such that 'm equals n' ($m = n$). Now under permutation this statement becomes 'n equals m' ($n = m$). But this is no change, since it is a well-known property of the equality relation that 'm equals n' implies 'n equals m' and vice versa. So the equality relation between two numbers possesses permutation symmetry (under interchange of the two numbers). This might seem trivial. Consider, then, another relation between two numbers, the relation of 'greater than' (>). The numbers m and n might be such that 'm is greater than n' ($m > n$). Permuting the numbers, we have 'n is greater than m' ($n > m$). These two statements contradict each other, so the 'greater than' relation does not possess permutation symmetry. (The same is obviously true of the 'lesser than' relation (<).)

PROBLEM

Find additional, diverse examples of permutation symmetry.

Group theory. The set of all permutations of n objects, including the identity permutation, with consecutive permutation as composition, forms a group, called the *permutation group* or *symmetric group* (not to be confused with *symmetry* group) on n objects and denoted S_n. The order of S_n is $n!$ ($= n(n-1) \ldots 2 \cdot 1$). S_1 is just the identity. S_2 is isomorphic with C_2 and with D_1. S_3 is isomorphic with D_3. Refer to Alexandroff (GRP, 1), Bell (GRP, 3), Grossmann and Magnus (GRP, 6), and Budden (GRP, 7). There is a theorem (Cayley's theorem) that any group of finite order n is isomorphic with a subgroup of S_n. This is connected with the fact that each row (or column) of a group table (such as the examples in chapter 2) is a permutation of the first row (or column) of the table. See Budden (GRP, 7).

Color symmetry

'Oh, you're not Piglet,' he said. 'I know Piglet well, and he's *quite* a different colour.'
(A.A. Milne: *Winnie-the-Pooh*)

When a system is colored in more than one color, we can apply a *color transformation* to it. This is an interchange of the various colors. For example, if a system has two colors, say black and white, the only color transformation applicable to it is interchange of the two colors; the image state has black wherever the initial state has white and vice versa (fig. 4.6). With larger numbers of colors the possibilities increase. Thus a color transformation is a kind of permutation.

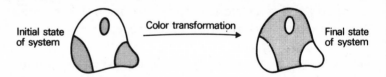

Fig. 4.6
Color transformation for 2-color system

No system can be invariant under color transformation, but there can be symmetry under the combination of color transformation with some other kind of transformation, where neither is a symmetry transformation by itself. Consider, for instance, the 1-dimensional pattern of fig. 3.3(*b*). A color transformation replaces black circles by white circles and white circles by black and is obviously not a symmetry transformation. A displacement by $\frac{1}{2}a$ (and odd multiples of it) in either direction gives the same result. But the combination of a color transformation and such a displacement (in either order, since they commute) is a symmetry transformation; each just cancels the effect of the other.

Since *color symmetry* cannot mean what we would expect it to mean, we use the term to denote symmetry under one or more transformations of which a color transformation is an ingredient. In this sense the system of fig. 3.3(*b*) possesses color symmetry. For 2-dimensional examples consider the lattices of figs. 3.8(*c*), 3.13(*c*), and 3.19(*c*). The addition of a color transformation allows new displacement, reflection and rotation transformations, not in themselves symmetry transformations, to become so. (See the bibliography under COL.)

PROBLEM

Find these transformations.

PROBLEM

Find all the geometric and color symmetries of the two M.C. Escher lattices of fig. 4.7. Refer to MacGillavry (ART, 6) for many additional color symmetric lattices by Escher.

Color symmetry of 3-dimensional lattices is very important, but we do not discuss it. (Refer to Shubnikov (COL, 1).) We do mention, though, that, in spite of our apparent preoccupation with lattices, color symmetry can be possessed also by finite systems, such as in fig. 4.8.

PROBLEM

Find all the geometric and color symmetries of the systems of fig. 4.8.

Group theory. The *color groups*, extensions of the point and space groups to include color transformations, are relatively recent discoveries. See under COL in the bibliography.

Analogy

'When you wake up in the morning, Pooh,' said Piglet at last, 'what's the first thing you say to yourself?'
    'What's for breakfast?' said Pooh. 'What do *you* say, Piglet?'
    'I say, I wonder what's going to happen exciting *to-day*?' said Piglet.
    Pooh nodded thoughtfully.
    'It's the same thing,' he said.
    (A.A. Milne: *Winnie-the-Pooh*)

An important, abstract kind of symmetry, that is rarely thought of as a symmetry, is *analogy*. This is the invariance of a relation or statement

Fig. 4.7
2-dimensional lattices with color symmetry by M.C. Escher. (Collection Haags Gemeentemuseum, The Hague)

(*a*) Circle

(*b*) Sphere (beach ball)

Fig. 4.8
Color symmetric finite systems

under changes of the elements involved in it. In the case of a relation between only two elements the analogy takes the form '*A* is to *B* as *C* is to *D*'. This means that the relation holding between *A* and *B* holds equally well between *C* and *D*, with *C* taking the place of *A* and *D* replacing *B*. Then the pair '*A*, *B*' and the pair '*C*, *D*' are said to be analogous with each other, and we say that there is an analogy between the pairs.

As an example, consider the relation between a circle and the plane in which it lies. A circle is the locus of points equidistant from a given point (its center), all points lying in a plane. The relation here is: *X* is the locus of points equidistant from a given point (its center), all points lying in *Y*. *X* and *Y* are the elements of the relation. For *X* = 'a circle' and *Y* = 'a plane' the relation holds. It does not hold, for instance, for *X* = 'a parabola' and *Y* = 'space'. But are there other '*X*, *Y*' pairs for which the relation does hold? One such pair is *X* = 'a sphere' and *Y* = 'space'. Indeed, a sphere is the locus of points equidistant from a given point (its center), all points lying in space (that is, not constrained at all). Thus a circle is to a plane as a sphere is to space. The pair 'circle, plane' and the pair 'sphere, space' are analogous with each other. Can the analogy be extended? What must *X* be, if we take *Y* = 'a line', for the relation to hold? We see that *X* = 'a pair of points' is the answer. A pair of points is the locus of points equidistant from a given point (its center), all points lying in a line. So a pair of points is to a line as a circle is to a plane as a sphere is to space. Now we have three analogous pairs (fig. 4.9). And, allowing ourselves the abstractive license of a mathematician, we can even push this to infinity by defining 'an *n*-sphere' as that *X* for which the relation holds with *Y* = '*n*-dimensional space'.

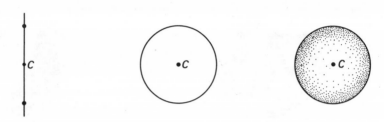

Fig. 4.9
Three analogous pairs: a pair of points in a line, a circle in a plane, and a sphere in space. Center is indicated for each by *C*

For another example return to the section on temporal symmetry at the beginning of this chapter. 'The period of temporal symmetry is analogous with the minimum displacement interval of geometric symmetry.'

An analogy might, of course, involve a relation among more than two elements or might involve a statement about a single element. In the latter case all the things, concepts, creatures, etc. for which the statement holds are analogous. For example, part of the preface of this book is based on an analogy between symmetry and a disease.

PROBLEM

Find additional analogies and consider them from the symmetry point of view.

Approximate symmetry

"Suppose *I* carried *my* family about with me in *my* pocket, how many pockets should I want?'
    'Sixteen,' said Piglet.
    'Seventeen, isn't it?' said Rabbit. 'And one more for a handkerchief – that's eighteen . . .'
    There was a long and thoughtful silence . . . and then Pooh, who had been frowning very hard for some minutes, said: '*I* make it fifteen.'
(A.A. Milne: *Winnie-the-Pooh*)

In our discussions of spatial and temporal displacement symmetries we introduced the concept of *approximate symmetry*. In a way this was forced upon us, because these symmetries are strictly applicable only to spatially or temporally infinite systems, and without recourse to approximate symmetry we would not be able to find examples of displacement symmetry. In those discussions we defined and considered

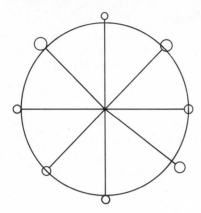

Fig. 4.10
Example of approximate symmetry

approximate *displacement* symmetry only. We now define the concept of approximate symmetry in full generality so as to allow its
application to any symmetry.

A system is said to possess approximate symmetry with respect to
a transformation, if it is almost invariant under that transformation.
The key word here, of course, is 'almost'. Just what 'almost' means
depends on the properties of the system, on the transformation under
consideration, and even on the personal taste of the observer, and must
be determined for each case. For example, if a system consists of
eight identical parts arranged regularly around an axis, it possesses
8-fold rotational symmetry. If the eight parts are not quite identical
or if their spatial arrangement around the axis is not quite regular, the
system can be said to have approximate 8-fold rotational symmetry
(fig. 4.10). It is then for the observer to decide how much of a deviation from identicalness or from regularity he is willing to allow and
still consider the situation to be one of approximate symmetry.

Another term used in this connection is *broken symmetry*. Broken
symmetry is approximate symmetry, but usually indicates special
interest in the deviation from exact symmetry, the *symmetry breaking*.
When the symmetry breaking can be considered to be brought about
by some cause, such that in the (possibly hypothetical) absence of this
cause the symmetry becomes exact, the cause is said to break the symmetry. The complete absence of such symmetry breaking, whether
obtainable in practice or not, is called the *exact symmetry limit*. An
example of a system with broken symmetry is a crystal, whose lattice

possesses approximate displacement symmetry. The crystal's displacement symmetry is approximate, and not exact, due to the crystal's finite size. So in this case it is the finiteness of the system that breaks the symmetry. The exact symmetry limit is an infinite crystal and is obviously unobtainable in practice.

PROBLEM

Find or invent various additional examples of broken symmetry. For each example clearly state what breaks the symmetry and define the exact symmetry limit.

5
Symmetry in nature

Nature is a wonderful source of symmetry. She exhibits symmetries of the kinds we discussed in the preceding chapters as well as others that we will describe in this chapter and more that we will not. Some of her symmetries are obvious, though she may often be coaxed into disclosing symmetries of amazing subtlety. Nature's symmetries appear on scales varying from the extremely large, the cosmological scale, to the extremely small, the subnuclear (elementary particle) scale. See Weyl (SYM, 1) and Kepler (PHYS, 5). Fundamental and relatively simple phenomena are often (though not always!) more symmetric than those that are compound and relatively complex. So it is the realm of physics, the study of nature at her most basic, that offers us the richest assortment of symmetries of nature. A general reference on modern views of the physical world is Gamow (SCI, 4). Also refer to Gamow (SCI, 5), Massey and Quinton (SCI, 6) and Feynman (PHYS, 14).

Let us start by reviewing some of the symmetries with which we have become acquainted so far in this book, with the purpose of showing by example how nature realizes those among them that she does in fact realize.

Symmetries with which we are acquainted

And then this Bear, Winnie-the-Pooh, F.O.P. (Friend of Piglet's), R.C. (Rabbit's Companion), P.D. (Pole Discoverer), E.C. and T.F. (Eeyore's Comforter and Tail-finder) — in fact, Pooh himself — said something so clever that Christopher Robin could only look at him with mouth open and eyes staring, wondering if this was really the Bear of Very Little Brain whom he had known and loved so long.
(A.A. Milne: *Winnie-the-Pooh*)

There are no material systems in nature that possess exact spatial displacement symmetry. But crystals make good approximations to such systems. As we mentioned before, crystals are finite 3-dimensional lattices constructed of ions, atoms or molecules. The smallest minimum displacement interval of a crystal lattice might typically be of the order of one millionth of a millimeter. So a crystal of millimeter size has a ratio of minimum displacement interval to total length of about one to a million, which is not a bad approximation to an infinite system for the present purpose. Examples of crystals are diamonds (crystallized carbon), rubies and sapphires (crystallized aluminum oxide), rock candy (crystallized sugar), ice (crystallized water), grains of table salt (sodium chloride crystals) and quartz crystals. The external appearance of crystals often (though not always!) reflects some or all of the symmetry of the lattice. References on crystals and crystallography are listed in the bibliography under CRYS, including cross-references.

Spatial displacement symmetry is also realized by nature in a very different and abstract manner. This realization is the homogeneity of space. What is meant by this is that all points of space are equivalent, as far as nature is concerned, in that the laws of nature are the same everywhere. Roughly speaking, the idea is that an experiment should give the same result irrespective of where it is performed. But this statement must be interpreted with care. A rock dropped at the top of Mount Everest will obviously not fall in the same manner as a rock dropped from a submerged submarine. But the physical laws governing the rock's behavior *are* the same at both locations. These laws take into consideration factors such as gravity, viscosity, pressure and temperature, and correctly predict the behavior of the rock wherever it is dropped. So everything else being equal, the rock's behavior is independent of its position in space. Whether space is exactly homogeneous, only approximately so, or not at all so for very large distances, is a cosmological question, and the general theory of relativity has much bearing on it. In addition, general relativity correctly predicts detectable deviations from homogeneity of space near very massive objects, such as the sun. But for most 'practical' purposes space can be considered exactly homogeneous. Thus assumption is one of the cornerstones of the special theory of relativity and is intimately related to the law of conservation of momentum. References to relativity are listed in the bibliography under REL, including cross-references. For conservation laws refer to listings under CONS and the cross-references.

Nature does not seem to be able to present us with an example of a material system with exact temporal displacement symmetry, which would necessarily be an eternally existing system. A possible exception is the universe itself, according to the 'steady state' cosmological theory. This theory suggests that the universe had no beginning in time and will have no end and that its general appearance is always the same on the average. But, as we mentioned earlier, present evidence seems to indicate that the evolution of the universe is better described by the 'big bang' theory than by the 'steady state' theory, so this possibility appears to be ruled out. Refer to entries in the bibliography under COS, including the cross-references.

On the other hand, nature abounds with examples of approximate temporal displacement symmetry. Such are all the natural periodic systems whose lifetimes are much longer than their periods. Astronomical examples are perhaps the most obvious, and several of these were presented in chapter 4. Additional examples are periodic biological processes, either linked to external cycles or independent of them. The sleep cycle of higher animals and the opening–closing cycle of flowers are linked to the diurnal cycle of day–night produced by the earth's rotation about its axis. The reproductive and migratory cycles of many higher animals as well as the leaf shedding and renewal cycle of deciduous plants are linked to the annual cycle of the revolution of the earth about the sun. The periodic process taking place in the heart which produces the heart's pumping action, the heartbeat, seems to be essentially independent of external cycles. The human female menstrual cycle is also practically independent of external cycles, though at some stage in evolutionary history it was probably linked to the lunar cycle of brighter and darker nights, just as the behavior of certain sea life is at the present time. Refer to Pengelley and Asmundson (TIM, 5), Schlegel (TIM, 6), Sollberger (TIM, 7), Luce (TIM, 10, 11) and Strughold (TIM, 12).

In analogy with homogeneity of space as a realization by nature of spatial displacement symmetry, temporal displacement symmetry is also realized by nature as homogeneity of time. This means that all moments of time are equivalent, in that the laws of nature are constant, and an experiment should give the same result whenever it is performed. (The last statement is subject to the same reservations as in the spatial case.) The question of whether time is exactly homogeneous, only approximately so, or not at all homogeneous over very large time intervals, is one of general relativity and cosmology. But, as in the spatial case, for 'practical' purposes time can be considered exactly homogeneous. This assumption is another of the cornerstones of special relativity and is intimately connected with the law of

(*a*) Globular star cluster (NGC 5272)

(*b*) Saturn

Fig. 5.1
Rotational symmetry of astronomical objects. (Hale Observatories photographs)

(*c*) Galaxy with arms (NGC 4565) seen edge on

(*d*) Galaxy with arms (NGC 1300)

conservation of energy. Refer to listings under REL and CONS in the bibliography, including cross-references.

Examples of many kinds of rotational symmetry are abundant in nature. As we mentioned previously, crystal lattices may possess rotational symmetry with respect to one or more variously oriented sets of parallel axes, but each set of symmetry axes is limited by the crystallographic restriction to being 2-fold, 3-fold, 4-fold or 6-fold. Astronomical objects tend to be spherical if they do not rotate, like globular star clusters (fig. 5.1(*a*)), but to have only axial symmetry if they rotate, like the planet Saturn (fig. 5.1(*b*)). Rotating galaxies tend to flatten out into discs with axial symmetry, while some go even further and develop a pair of arms, which reduces their rotational symmetry from axial to 2-fold (fig. 5.1(*c*), (*d*)). See the bibliography under CRYS and COS.

The external appearance of plants and animals often displays approximate rotational symmetry. Starfish have a single axis of 5-fold rotational symmetry (fig. 5.2(*a*)). Many flowers show rotational symmetry (fig. 5.2(*b*), (*c*)). The skeletons of radiolarians have several axes of rotational symmetry, and some have the rotational symmetry of a Platonic solid (fig. 5.2(*d*)). (See the discussion of rotational symmetry in the section on spatial symmetry of chapter 3.) Refer to Thompson (LIV, 7) and Horne (LIV, 8).

Snowflakes generally possess a single axis of 6-fold rotational symmetry (fig. 5.3). See Bentley and Humphreys (CRYS, 1) and Kepler (PHYS, 5).

The four bonds that a carbon atom usually makes with other atoms in the formation of molecules are, unless distorted, symmetrically oriented in space, so that, if the carbon atom is imagined at the center of a regular tetrahedron, each bond is directed toward the center of one of the four faces (fig. 5.4). Refer, for instance, to Pauling and Hayward (PHYS, 3). Thus each bond lies on an axis of 3-fold rotational symmetry. (The rotational symmetry of the regular tetrahedron was discussed in the section on spatial symmetry of chapter 3.) This fact is essential to organic chemistry, and so much so that a journal devoted to the subject is named *Tetrahedron*. When carbon atoms combine with each other in such a way that all bonds are utilized and none are distorted, the result is normally the diamond lattice (fig. 5.5).

At the subnuclear scale spinless particles, such as pions (pi-mesons) and kaons (K-mesons), have spherically symmetric properties, while the properties of particles with spin, such as electrons, positrons,

(*a*) Starfish

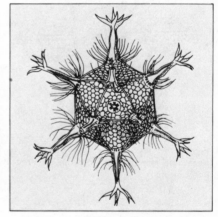

(*b*) Tickseed (coreopsis) flower

(*c*) Chrysanthemum flower

(*d*) Skeleton of radiolarian *Circogonia icosa-hedra* (from Haeckel: *Challenger Monograph*)

Fig. 5.2.
Rotational symmetry of plants and animals

protons and rho-mesons, exhibit axial symmetry. Refer to Ford (PART, 4), Frisch and Thorndike (PART, 6), Yang (PART, 11) and Gouiran (PART, 14).

Rotational symmetry is also realized abstractly by nature. So in addition to being homogeneous, as the realization of nature's spatial displacement symmetry, space is also isotropic. This means that, as

Fig. 5.3
Rotational symmetry of snowflakes. (From Bentley and Humphreys: *Snow Crystals* (Dover Publications))

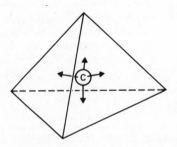

Fig. 5.4
Tetrahedral spatial orientation of carbon bonds

far as nature is concerned, all directions are equivalent, and the laws of nature are independent of orientation. An experiment performed at a point in space should (with the reservations explained in the

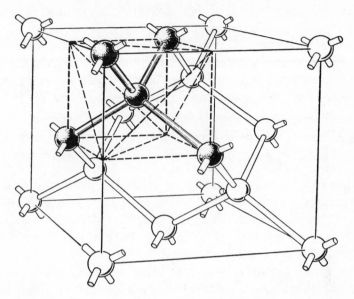

Fig. 5.5
Diamond lattice. Sample tetrahedron is shown in small cube. (Adapted from
Shockley: *Electrons and Holes in Semiconductors* (Van Nostrand Reinhold
Co., copyright 1950 by Litton Educational Publishing, Inc.))

homogeneity case) give the same result regardless of the orientation of
the apparatus. Whether space is exactly isotropic or not is again a
matter for cosmology and the general theory of relativity. But for
'practical' purposes it is safe to assume that it is. This is another basic
assumption of the special theory of relativity. It is intimately related
to the law of conservation of angular momentum. Refer to the bibliography under REL and COS, including cross-references.

Spatial reflection symmetry is quite common in nature. Crystal
lattices may possess, in addition to various combinations of different
kinds of rotational symmetry, various combinations of reflection symmetries together with them. There are, in fact, 230 distinct combinations. These are the crystallographic space groups mentioned in
the section on spatial symmetry of chapter 3. These 230 kinds of
lattice symmetry serve to classify crystals, since any given crystal can
possess one and only one such symmetry. As noted before, these
symmetries are grouped into 32 crystal classes, which are then further
divided into seven (sometimes six) crystal systems. Refer to Weyl
(SYM, 1), Coxeter (GEO, 1) and listings under CRYS in the bibliography.

Another important appearance of reflection symmetry in nature is in plants and animals, especially the possession of a single plane of reflection symmetry. This symmetry is called *bilateral symmetry* by botanists and zoologists and is most common among the so-called higher plants and animals, that is, those species considered to be more advanced from the evolutionary point of view. Mammals, birds, reptiles, amphibians and fish are typically bilaterally symmetric. Leaves often have bilateral symmetry. Plants whose flowers are bilaterally symmetric are considered evolutionally more advanced than those producing rotationally symmetric ones (fig. 5.6).

However, this bilateral symmetry in higher plants and animals is not exact, certainly not from the internal anatomical and physiological aspect, but not even in external appearance. In humans the positions of the heart and the appendix and the shape of the stomach, for example, break bilateral symmetry. And it has been shown that the functions of the two halves of the brain are not symmetrically related. In external appearance, although the human form is grossly bilaterally symmetric, there are always 'defects' that break the symmetry. For example, the fingerprints of one hand are not the mirror images of the respective fingerprints of the other hand. And the hand and foot of one side (usually the right side for right handed people) are almost always slightly longer than those of the other side. Refer to Corballis and Beale (LIV, 3), Gazzaniga (LIV, 4), Sperry (LIV, 6), Kimura (LIV, 10), Thompson (LIV, 7), Weyl (SYM, 1), Gardner (REF, 1) and Fritsch (REF, 2).

An additional symmetric aspect of plants is *phyllotaxis*, the arrangement of leaves about a stem or the arrangement of other parts of the plant. Reflection symmetry is not necessarily involved here (for example, screw symmetry is applicable to the arrangement of leaves about a stem), and we mention the subject at this point only because we were discussing plants. We will not go into further details, but suggest that the interested reader investigate on his own. Refer to Church (LIV, 1, 2), Snow (LIV, 5), Thompson (LIV, 7), Weyl (SYM, 1), Coxeter (GEO, 1) and Jaeger (PHYS, 4).

Does nature exhibit reflection symmetry in the abstract manner that we described for spatial displacement symmetry, temporal displacement symmetry and rotational symmetry? Yes and no. The situation in the case of reflection symmetry is not as simple as for the other symmetries. To be able to explain this, we must first make clear just what reflection symmetry of nature means. In analogy with the

(*a*) Arum leaf

(*b*) Orchid flower

Lionfish

(*d*) Jay

Fig. 5.6
Bilateral symmetry of plants and animals

(*e*) Human and teddy bear

meaning of the other symmetries as symmetries of nature, reflection symmetry of nature is the equivalence of all systems with their reflection images, in that the laws of nature are the same for both. In other words, if an experiment gives a certain result, then the result of another experiment, related to the former by a reflection transformation, should be the reflection image of the former result. (Here we are concerned with real reflections, that is, plane or point reflections, and not line reflections, which are equivalent to rotations. Due to isotropy of space (the rotational symmetry of nature), it makes no difference whether we consider here plane or point reflections, since they only differ by a rotation. Refer to the end of the discussion of spatial reflections in chapter 3.)

Now it is known that not all laws of nature apply equally to systems and to their reflection images. The exceptions appear in subnuclear phenomena and have to do with the weak interaction, one of the forces among elementary particles. The weak interaction is responsible for, among other effects, the beta decay of various radioactive isotopes, whereby a neutron in the nucleus spontaneously transmutes into a proton, which remains within the nucleus, and an electron ('beta particle') and antineutrino, which leave the nucleus. Refer to Trieman (PART, 9), Ford (PART, 4), Frisch and Thorndike (PART, 6), Gouiran (PART, 14), Gamow (SCI, 4), and Feynman, Leighton and Sands (PHYS, 13). Another of the forces by which elementary particles act on each other is the strong interaction, whose most important function in nature is to hold the components of nuclei together. The strong interaction is, as far as we presently know, reflection symmetric. Refer to Ford (PART, 4), Frisch and Thorndike (PART, 6) and Gouiran (PART, 14).

Elementary particles carrying electric charge, such as the proton and the electron, also affect each other through the electromagnetic interaction, which appears as the electric force ('like charges repel each other; opposite charges attract each other') and the magnetic force. (Electrically neutral particles, such as the neutron, can also interact via the electromagnetic interaction with each other and with charged particles for reasons that we cannot go into here.) The electromagnetic interaction has a disruptive effect on nuclei, due to the mutual repulsion of the protons in the nuclei, but this is overcome by the strong interaction. See Gamow (SCI, 4). On the other hand, the electromagnetic interaction binds electrons into orbits around nuclei to form atoms and allows atoms to combine to form molecules. It supplies the force that holds crystals and other solids together, furnishes the weaker attraction among molecules that produces liquids, and bounces the molecules of gases about. The electromagnetic interaction is the force that acts on the level of atomic and molecular physics, chemistry, biology and everyday affairs. It has reflection symmetry. The only other force that we know of in nature is the gravitational force. It appears to be insignificant as a force among individual particles, atoms or molecules, but becomes important at larger scales, and is practically the only force acting at the astronomical and cosmological levels. Again, as far as we know, it possesses reflection symmetry. Refer to Feinberg (SCI, 2) and Hughes (PART, 13).

Thus all the laws of nature responsible for the existence of bulk matter and its properties, that influence our everyday lives, are reflection symmetric. And in this sense we claim that nature possesses reflection symmetry. But the weak interaction, though it hardly has practical effect on these levels, *is* an asymmetric aspect of nature. So nature is not strictly reflection symmetric. We might say that she exhibits broken reflection symmetry. Conservation of parity is closely related to reflection ·symmetry. Concerning reflection symmetry, its breaking and parity, refer to Gardner (REF, 1), Lee (PART, 8), Wigner (PART, 10), Yang (PART, 11), Ford (PART, 4), Frisch and Thorndike (PART, 6), Gouiran (PART, 14), Feynman, Leighton and Sands (PHYS, 13) and Feynman (PHYS, 14).

We now come to temporal reflection symmetry. Physicists usually call this transformation time reversal, and we will too. Nature offers very few examples of systems whose development exhibits time reversal symmetry. Among these are systems undergoing simple harmonic linear vibration, for instance diatomic molecules, such as H_2, N_2, CO. As for the question of whether nature possesses time reversal symmetry in the abstract way we discussed for other symmetries, like in the case of spatial reflection symmetry the answer depends on which aspect of nature is being considered. The meaning of this kind of time reversal symmetry is that the laws of nature are the same for processes as for their temporal reflections. That is, if a motion picture of any process allowed by nature is run backward, the result is also a process allowed by nature.

Starting with nature on her very largest scale, the scale of the universe, we already run into trouble. The universe is expanding, and the time reversal of this process is a contracting universe. Is a contracting universe allowed by the laws of nature? This is a puzzle, because there is only a single universe and it is expanding. We cannot show by example that a contracting universe is allowed. And we do not know enough about cosmology to be able to state whether the universe has ever contracted or will ever contract, or whether it could theoretically contract but now happens to be expanding, or whether the latter statement even means anything. See under COS in the bibliography, including cross-references, and especially Bonnor (COS, 19).

If we consider nature on smaller scales, though, more definite answers can be given. On the astronomical scale some phenomena exhibit approximate time reversal symmetry, while others lack it. On the scale of everyday objects and events there is no time reversal symmetry. Exact symmetry is found on the subnuclear level. The first point to make clear is that there is no time reversal symmetry for systems whose behavior is affected by their being composed of large numbers of constituents. Such systems have statistical properties and are therefore subject to the laws of thermodynamics. The second law of thermodynamics is what interests us here, and it states that the entropy, a measure of the degree of disorder, of a system will never decrease as the system develops in time. In reality the entropy of systems (and their disorder) is continually increasing; development with constant entropy is a theoretical ideal. Time reversal of a process with increasing entropy gives as the image a process with decreasing entropy, which is prohibited by the second law of thermodynamics. Therefore any time reversal symmetry that nature might have cannot be realized by means of such systems.

To illustrate this point consider two containers connected through a valve, one filled with hot water and the other with cold (fig. 5.7). The valve is opened and the hot and cold water mix, the final result being two containers filled with warm water. The initial state has more order, and thus lower entropy, than the final state, because the H_2O molecules are initially sorted, those in the hot water having on the average more thermal energy than those in the cold, while in the final state this sorting is lost. The time reversed process, which is not known to have ever occured, is the spontaneous conversion of the two containers filled with warm water to one container of hot water and one of cold. For this to happen the molecules must 'voluntarily' segregate themselves, with the more energetic ones collecting in one container and the less energetic in the other, thus increasing the degree of their order and so decreasing the entropy of the system. Such a process would be in contradiction with our experience and would violate the

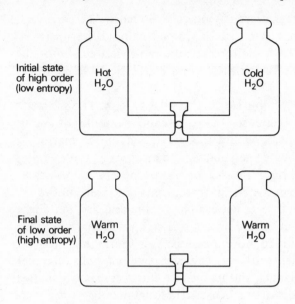

Fig. 5.7
Irreversible mixing as example of time reversal asymmetric process

second law of thermodynamics, which is just a precise formulation of our experience. Refer to Gardner (TIM, 2), Ehrenberg (TIM, 1), Schlegel (TIM, 6), Feynman, Leighton and Sands (PHYS, 13), Feynman (PHYS, 14) and Bent and Bent (TIM, 13).

On the everyday scale of objects and events systems are strongly affected by their being composed of large numbers of constituents – ions, atoms, molecules. Irreversible mixing, as in the example, contributes to the disorder. Friction and viscosity cause orderly motion to degenerate into random thermal motion. This is why we are deprived of ordinary sized realizations of any time reversal symmetry nature might have.

The second point is that the nonstatistical basic laws of nature, responsible for the existence of bulk matter and its properties, possess time reversal symmetry. The strong interaction holding the nucleus together has this symmetry, as far as we know. The electromagnetic interaction, responsible for forming atoms, molecules and conglomerations of these, is also time reversal symmetric. The gravitational interaction holds matter together on the astronomical scale and seems to be symmetric on this scale, though its properties on the cosmological scale are less clear. And the laws of mechanics, Isaac Newton's three laws, definitely have time reversal symmetry. So nature's time reversal symmetry can be realized by means of systems whose behavior is governed by the above laws and interactions and that are composed of only a small number of constituents or are effectively so composed. See Gardner (TIM, 2).

The last situation is exemplified by the solar system. The sun and its planets are certainly composed of a tremendous number of components — molecules, atoms, ions. But to a very good approximation this fact can be ignored when calculating the motions of these bodies. They can be considered point masses for this purpose. As a result, a time reversed solar system, where all revolutions take place in the opposite direction from the actual, is a physical possibility and does not contradict any law of nature. But this *is* an approximation. When the solar system is examined more precisely, it is found that the structure of the planets comes into play. The ever changing pulls on a planet by the sun, the other planets, and its own satellites constantly distort its shape. This varying distortion causes frictional heating of the planet, which is paid for at the expense of orbital and rotational energy of the planets and their satellites. Thus the second law of thermodynamics takes its toll, as the solar system very gradually converts its orderly energy into random thermal motion. This is the kind of astronomical phenomenon we referred to above as exhibiting approximate time reversal symmetry. Effectively (and to a good approximation) the system contains only a small number of constituents.

Examples of systems which really consist of small numbers of constituents are systems of small numbers of elementary particles. (To be more precise, it is still an open question whether the so-called 'elementary' particles are in fact the ultimate constituents of matter or not. But, at least, the possibility that they are is not excluded by our present knowledge.) The time reversed images of elementary particle processes seem to be allowed by nature and obey the same laws as the direct processes, whether these processes are governed by the electromagnetic, the strong or the weak interaction. So nature can exhibit her time reversal symmetry on this scale. (A possible reservation is discussed in the next section of this chapter.) Refer to Overseth (TIM, 4), Gardner (TIM, 2), Wigner (PART, 10), Lee (PART, 8), Ford (PART, 4), Frisch and Thorndike (PART, 6), Gouiran (PART, 14) and Feynman, Leighton and Sands (PHYS, 13).

A curious example of approximate dilation symmetry in nature is Bode's law. This is the observation that the distances from the sun of the planets of the solar system, including the asteroid belt, form a geometric series to a good approximation. The distance of each planet from the sun is about $\sqrt{3} = 1.73$... times that of its inner neighbor. In fact, with simple modifications the law not only fits the planetary distances closely, but is also applicable to satellite systems of planets. Bode's law reminds us of the example of dilation symmetry involving concentric circles that we showed in chapter 3 (fig. 3.37), with elliptical orbits instead of circles, semi-major axes replacing radii, a factor of $\sqrt{3}$ instead of 2, and a finite number (ten) of orbits instead of infinity. So we have here a system possessing approximate dilation symmetry, approximate due both to the imperfect fit to a geometric series and to the finiteness of the system. Just how far does nature carry this symmetry? Are there additional, as yet undiscovered planets extending the Bode series, liquid blobs inside Mercury's orbit or lonely travelers out beyond Pluto? Only observation will tell. Refer especially to Nieto (COS, 21).

Concerning dilation symmetry the interesting question is whether, by imposing the same laws on systems differing only by dilation transformations, nature exhibits this symmetry. The special theory of relativity requires that, if such a symmetry exists at all, the dilation transformations must affect time and space equally. So the transformations we are now considering apply also to the temporal development of systems and change all time intervals in this development by the same factor that they change all lengths in the developing system. For example, if all lengths are doubled, then all time intervals in the development are also doubled, making the development progress at half the speed. The question of nature's dilation symmetry can be put as follows: given a system developing according to the laws of nature, are all developing systems, differing from this one by dilation transformations only, consistent with the same laws of nature? And the answer is negative. For reasons that we cannot go into here, such space–time dilation transformations as are being considered must also change all masses of the system by a factor equaling the reciprocal of the factor by which the spatial and temporal intervals are changed. Now the kinds of elementary particles found in nature are an aspect of the laws of nature. Nature allows, in fact requires, the existence of these kinds of particles and forbids the existence of other kinds. (Physicists do not yet understand at all why this is so.) And one of the properties characterizing each kind of elementary particle is its mass. If nature were dilation symmetric, each kind of particle would have to appear in an infinity of versions, having all possible masses, but otherwise identical. (Only the zero-mass particles, which are the photon and the two kinds of neutrinos and antineutrinos, would be exempt from this proliferation. Concerning the two kinds of neutrinos, see Lederman (PART, 7).) This is clearly not the situation in reality. So nature does not possess dilation symmetry on this level, the most fundamental, even though in certain circumstances she might exhibit approximate symmetry on this or on larger scales. See Feynman, Leighton and Sands (PHYS, 13) and Feynman (PHYS, 14).

The appearance of permutation symmetry in nature requires the existence of a set of identical objects. On the medium and large scales there are no identical objects; there are similar objects, but no strictly identical ones. Identical objects are found only on the elementary particle level. All particles of the same kind, such as all electrons or all protons, are identical. There is no way of distinguishing one from another. Elementary particles cannot be individually labeled. So a system containing, say, two electrons among its constituents (such as a helium atom) is symmetric under interchange of its electrons.

This type of symmetry is not just a subject for idle observation, but has far reaching consequences. It is intimately related to Wolfgang Pauli's exclusion principle, which states that no two electrons can be in the same state, that is, can have exactly the same momentum and direction of spin axis. (This is valid also for two of any other kind of particle characterized by spin of magnitude $\frac{1}{2}$ unit, such as protons and neutrons, or $\frac{3}{2}$ unit, $\frac{5}{2}$ unit, etc.) Refer to Gamow (EXC, 1). The Pauli principle is fundamental in explaining the chemical properties of elements and is part of P.A.M. Dirac's theory of positrons (antielectrons) as 'holes in the infinite sea of negative-energy electrons'. Refer to Gamow (EXC, 1) and (SCI, 4). Another consequence of this permutation symmetry is that the statistical distribution of energy among the particles of elementary particle 'gases' is different from the energy distribution among the molecules of ordinary gases, and this affects various properties of these 'gases'. Examples are the conduction electrons (those that conduct electricity) in metals, which behave like a gas of electrons confined to the volume of the metal, and neutron stars, which are extremely dense stars consisting solely of neutrons. See Gamow (EXC, 1), Mott (CRYS, 4), Ostriker (COS, 11) and Ruderman (COS, 14).

Color symmetry as such does not seem to be of much importance in nature. But the theory of color symmetry is very important for the study of the physical properties of crystals, where different colors represent the different kinds of ions, atoms or molecules comprising the crystal. Refer to Holser (COL, 1) and MacGillavry (ART, 6).

Analogy is abundant in nature. It can be found at all levels and in all fields of investigation. For example, Newton's law of gravitational force and C.A. de Coulomb's law of electrostatic force are analogous in that both forces act along the line connecting the interacting bodies and both are inversely proportional to the square of the distance between the bodies. This analogy gives rise to another analogy at a different level. An atom, a number of electrons revolving around a nucleus, is analogous with the planetary system of a star or the satellite system of a planet. Or all those mechanical, electromagnetic, electromechanical, geophysical, chemical, electrochemical, physiological, biological, etc. etc. systems that, when displaced slightly from a state of stable equilibrium, perform damped harmonic oscillations in their return to equilibrium are thus analogous with each other. Some examples in nature are: a lake or bay, whose water, when displaced from equilibrium by an earthquake or strong wind, oscillates back to equilibrium by sloshing back and forth within its confinement; the earth, which, when deformed by an earthquake, oscillates in its return to normal shape; a tree, that, when bent and released, might oscillate a few times before coming to rest.

Analogy is essential in thought and learning processes, including scientific discovery, and in education. For instance, it so often happens that an infant, upon learning that 'daddy' (or 'abba' or 'papa' etc.) is an appropriate utterance for one of those big moving holding feeding talking things, immediately and with excellent analogy applies it to the other thing. And when the little genius finally gets 'daddy' and 'mummy' differentiated in his mind, he causes some embarrassment, but uses good analogy, when he applies 'daddy' to all the men passing by in the street. Later he will immediately know the plural form of almost every new noun he learns without actually having heard it (if indeed his language has nouns and plural forms). He does this by analogy. After slowly learning many examples like toy-toys, finger-fingers, spoon-spoons, he realizes the analogy and forms the relation $X-X$s, which he applies from then on. The analogy will produce also mouses and foots, but the exceptions to the rule are soon mastered.

Scientific discovery would be impossible without analogy. Where would we be today, if each planetary orbit in the solar system were considered a thing in itself and no relation were perceived among the orbits? As it was, all the orbits were in fact seen to be analogous in that they all obeyed Johannes Kepler's laws. From these relations Newton arrived at his law of gravitation. See, for instance, Gamow (SCI, 4).

Symmetries we have not yet studied

'Here — we — are,' said Rabbit very slowly and carefully, 'all — of — us, and then, suddenly, we wake up one morning, and what do we find? We find a Strange Animal among us. An animal of whom we had never even heard before! An animal who carries her family about with her in her pocket!'
(A.A. Milne: *Winnie-the-Pooh*)

We now turn to symmetries of nature that have not been described in the preceding chapters. The first of these is connected with Albert Einstein's special theory of relativity, which has already been mentioned in this chapter as being based in part on various symmetries of nature — homogeneity and isotropy of space and homogeneity of time. For the present discussion let us reformulate these symmetries in terms of two well-equipped experimental physicists out to check the laws of nature. Homogeneity of space means that, if the two physicists perform experiments at different locations but at the same time and with their apparatuses oriented in the same direction, they will find the same laws of nature. Isotropy of space implies that they will find the same laws of nature, if they perform their experiments at the same location and at the same time but with their apparatuses oriented differently. Homogeneity of time states that, if the two physicists check the laws of nature at the same location and orientation but at different times (actually a single physicist could do this), they will also find the same laws of nature.

In all these cases it is tacitly assumed that both physicists are at rest. The most important of the basic assumptions of special relativity has to do with physicists in relative motion. It claims that, if one physicist is at rest and the other moves at constant speed in a straight line, they will still find the same laws of nature. This can be formulated as a symmetry by stating that the laws of nature are invariant under changes of the constant straight-line velocity of the laboratory in which the laws are tested. These transformations are called *velocity transformations* or *boosts*. (Since we cannot devote an excessive amount of space to special relativity, we purposely refrain from discussing important details, such as: With respect to what are rest and constant straight-line velocity determined? How is invariance of laws of nature expressed?) One such law is the speed of light $c = 299\,793$ kilometers per second, with the seemingly paradoxical result that every measuring device, whatever its velocity, will find this same value for c. Some of the better known correct predictions of the special theory of relativity are: the equivalence of mass m and energy E through Einstein's famous relation $E = mc^2$; the contraction of a body moving with velocity v by the factor $\sqrt{(1 - v^2/c^2)}$; the slowing down of a clock moving with velocity v by the same factor; the impossibility of accelerating a body to velocities equaling or exceeding the speed of light. Concerning special relativity, refer to Bondi (REL, 2), Gardner (REL, 3), Durell (REL, 4), Landau and Rumer (REL, 5), Taylor and Wheeler, (REL, 7), Gamow (SCI, 3, 4), Einstein (REL, 9) and Russell (REL, 10).

Group theory. The set of transformations under which the laws of nature are invariant, according to the special theory of relativity, form an infinite order group, with consecutive transformation as composition. This symmetry group of special relativity, called the *Poincaré group*, consists of the identity transformation, the velocity transformations (up to the speed of light), all rotations about a point, all spatial displacements, and all temporal displacements. Excluding the displacement transformations, we have a useful subgroup, called the *Lorentz group*.

Another symmetric aspect of nature is the roles played by electricity and magnetism in electromagnetic phenomena. At first glance this is not apparent. Electric effects are based on the electrostatic force between two electric charges, whereby like charges repel each other and opposite charges are mutually attracted. The electrostatic force acts parallel to the line connecting the two charges, and its magnitude depends only on the magnitudes of the charges and on the distance between them. The magnetic force, on the other hand, although it is also a mutual influence of two electric charges, does not act parallel to the line joining the charges, and its magnitude depends on the relative velocity of the two charges as well as on their magnitudes and separation. Every electric charge, whether at rest or in motion, produces an electric field, while only moving charges, that is, electric currents, produce magnetic fields.

To compute the fields produced by any given arrangement of charges and currents, one must solve J.C. Maxwell's equations, which are a set of equations with terms containing the electric field, the magnetic field, the electric charges, and the electric currents. These equations come in pairs, the number of pairs depending on the notation used, where both members of a pair are very similar. When the electric field appears in one member of a pair, the magnetic field appears in the other member at the corresponding position; and when the magnetic field appears in one, the electric field appears analogously in the other. But when the electric charges or currents appear in one member of such a pair, the other member contains a zero at the analogous position. This is because no *magnetic* charges have ever been discovered.

In the extremely compact notation of special relativity this should be most evident. Then Maxwell's equations take the form of a single pair,

$$\partial_\mu F^{\mu\nu} = \frac{4\pi}{c} J^\nu,$$

$$\partial_\mu F^{\dagger\mu\nu} = 0.$$

The symbol $F^{\mu\nu}$ stands for the various components of the electric and magnetic fields. $F^{\dagger\mu\nu}$ does also, but with the electric and magnetic fields interchanged. ∂_μ denotes derivatives with respect to the space–time coordinates. c is the speed of light. J^ν refers to the electric charge and current in the first equation; the lack of a corresponding term in the second equation, where a zero appears on the right hand side, reflects the nonexistence of magnetic charges and currents in nature, at least as far as is presently known.

If such magnetic charges, usually called magnetic monopoles, did exist, they would produce magnetic fields, just as electric charges produce electric fields. And moving magnetic monopoles (magnetic currents) would produce electric fields, just as moving electric charges (electric currents) produce magnetic fields. Magnetic monopoles would mutually attract or repel magnetostatically, and so on. Then the symmetry of nature under interchange of electricity and magnetism would be obvious. There are physicists who find this symmetry so compelling, and those zeros in Maxwell's equations so displeasing, that the experimentalists among them are actively searching for magnetic monopoles in nature, while the theoreticians engage in predicting their properties. See Ford (PART, 3).

An additional symmetry of nature, still within the realm of electromagnetism, is the invariance of the laws of electromagnetism under interchange of positive and negative electric charge. We are not referring to the arbitrariness in the assignment of signs to charges, that is, to the fact that it is unimportant whether the charge of the electron is called negative or positive as long as the charge of the proton is called the opposite, and so on with the other elementary particles. The point is that the electromagnetic behavior of any system of electric charges and currents is the same as that of a system identical with it except that all charges have the opposite sign. This is immediately clear from the electrostatic force. Like charges remain like charges after reversal of sign and continue to repel each other. Opposite charges remain opposite and still attract. And similarly for the magnetic force. This transformation of reversing the signs of all electric charges is called *charge conjugation.* So the electromagnetic interaction possesses charge conjugation symmetry.

Nature does not seem to discriminate basically between positive and negative electric charges. In fact, as far as is known, for every type of elementary particle nature presents a similar type, called the *antiparticle* of the former, having opposite electric charge. (An antiparticle is often called by the same name as the corresponding particle, with the addition of the prefix 'anti-'. For example, antiproton, antiomega-minus. But the antielectron is called positron.) Now particles are characterized by more than just their electric charge. And some of these additional properties have a charge-like character; the amount of each one that a particle carries is always a multiple of some unit and can be positive, negative or zero, and the total amount for a collection of particles is the sum of the amounts carried by the individual particles (with positive and negative canceling). Such 'charges' are: baryon number, electronic lepton number, muonic lepton number, third component of isospin and hypercharge. (Strangeness is related to some of these.) Particles and their antiparticles are not only opposite in the sign of their electric charge, but also in the signs of all their other 'charges' as well. On the other hand, they are identical in properties such as their mass and spin.

Several examples: The proton has electric charge $+1$, baryon number $+1$, lepton numbers 0, third component of isospin $+\frac{1}{2}$ and hypercharge $+1$. The antiproton has for these same properties the corresponding values of -1, -1, 0, $-\frac{1}{2}$ and -1. The electron has electric charge -1, baryon number 0, electronic lepton number $+1$ and muonic lepton number 0. Isospin and hypercharge are not defined for the electron. The positron has the corresponding values $+1$, 0, -1 and 0. Electrically neutral particles that are neutral (that is, have the value zero) for all their other 'charges' also, such as the photon and the neutral pion (pi-meson), are their own antiparticles. But the neutron, for example, even though it is electrically neutral, has nonzero values for others of its 'charges' and thus is distinct from its antiparticle. Refer to Feinberg and Goldhaber (CONS, 1), Lederman (PART, 7), Ford (PART, 4), Frisch and Thorndike (PART, 6), Gouiran (PART, 14), Feynman, Leighton and Sands (PHYS, 13), Feynman (PHYS, 14), and Hall and Mena (ANTI, 3). On the possible importance of antimatter (matter consisting of the antiparticles of the particles comprising ordinary matter) in cosmology see Alfvén (ANTI, 1, 2).

It is very useful to generalize the transformation of charge conjugation from the interchange only of positive and negative electric charges to particle–antiparticle interchange and thus to the interchange of positive and negative values for all the other 'charges' as well. This generalization is the transformation that physicists mean when they refer to charge conjugation. In fact, some people prefer to call it by the more precise, but longer, name of *particle–antiparticle conjugation*, and we do the same to avoid confusion. As far as the electromagnetic interaction is concerned, there is no difference between the two conjugations. This interaction depends only on the value of the electric charge of the particle and is independent of all the other 'charges'. But they are very different with respect to the strong and weak interactions, and it is the particle–antiparticle conjugation transformation that is of interest.

The strong interaction is invariant under particle–antiparticle conjugation. This means that the development of a system of particles interacting through the strong interaction is the same as that of a system identical to it but particle–antiparticle conjugated. So both the strong and the electromagnetic interactions have particle–antiparticle conjugation symmetry. And since, as we explained previously, these interactions are responsible for the existence and behavior of bulk matter, we expect bulk antimatter, should we ever find or produce any to observe, to behave in the same way with which we are familiar for bulk matter.

The weak interaction, on the other hand, is not particle–antiparticle conjugation invariant. The development of a system of particles interacting through the weak interaction is not in general the same as that of a system identical to it but with particles and antiparticles interchanged. And, as we mentioned previously, neither is the weak interaction symmetric under spatial reflection. So the weak interaction would seem to be a hopeless case symmetrywise. But it is not. It still exhibits a very subtle symmetry. It is symmetric with respect to the combined transformation of particle–antiparticle conjugation and spatial reflection. That is, the result of a weak interaction experiment, which is the spatial reflection and particle–antiparticle conjugation image of another experiment, is the spatial reflection and particle–antiparticle conjugation image of the result of the other experiment. In picturesque language, the symmetry mirror for the weak interaction also 'reflects' particles into their antiparticles. Refer to Wigner (PART, 10), Lee (PART, 8), Gardner (REF, 1). Yang (PART, 11), Ford (PART, 4), Frisch and Thorndike (PART, 6), Gouiran (PART, 14) and Feynman, Leighton and Sands (PHYS, 13).

For the continuation of this discussion it is useful to introduce abbreviations for three transformations. We denote spatial reflection by P (for parity), particle–antiparticle conjugation by C, and time reversal by T. This is standard notation among elementary particle physicists. The combined transformation of any two of these or all three is indicated by writing the appropriate symbols together (in any order, since these transformations commute with each other). For example, we just mentioned that the weak interaction is not symmetric under P or C separately, but that it is CP symmetric. The strong and electromagnetic interactions are both P symmetric and C symmetric, while all three interactions appear to have T symmetry.

There are compelling theoretical reasons (which we cannot go into here) to believe that the combined transformation CPT must be a symmetry transformation for any elementary particle interaction. This means that, regardless of the symmetry or asymmetry of an interaction with respect to C, P and T separately and to the pairs CP, CT and PT, the combination CPT must be a symmetry transformation for the interaction. For any elementary particle process allowed by nature the time-reversed, particle–antiparticle conjugated, spatially reflected process must also be allowed by nature. This clearly holds for the strong and electromagnetic interactions, since they are both separately symmetric under P, C and T, and thus also under CPT. The weak interaction is not symmetric under C and P separately, but is symmetric under T and under the combination CP, and so it too has CPT symmetry. The gravitational interaction is too weak to investigate on this level at present.

However, all is not so well. What appears to be a violation of CP symmetry has been observed in a process involving the neutral kaon (K-meson). Theorists have not yet reached agreement on an explanation for this, whether the CP symmetry violation is due to a property, unknown until now, of the familiar interactions or to a new, superweak interaction. In any case, if CP symmetry does not hold for a process, then, if *CPT* symmetry is to remain valid, neither can *T* be a symmetry transformation. *T* asymmetry is clearly needed to cancel out the *CP* asymmetry and thus to allow *CPT* symmetry. Such an effect is being searched for experimentally. But if it is not found and *CPT* symmetry is shown to be invalid, a considerable revision of certain concepts in theoretical physics will become necessary. Refer to Wigner (PART, 10), Lee (PART, 8) and Gardner (TIM, 2).

The realm of elementary particles displays many additional interesting symmetries, called *internal symmetries*, which are of an abstract nature. They cannot be described in terms of transformations in space or time, but are expressed as transformations in abstract mathematical 'spaces'. We do not attempt to explain these symmetries further, but just present a list of the more firmly established ones. There is an internal symmetry associated with each of the conservations of electric charge, baryon number, electronic lepton number and muonic lepton number. The strong interaction exhibits (broken) SU(3) (also called 'eightfold way') symmetry, which includes symmetries connected with conservation of total isospin and of hypercharge. The electromagnetic interaction has internal symmetries related to the conservation of hypercharge and of the third component of isospin. Theoretical physicists are still trying to understand this complex of internal symmetries and its relation, if there is any at all, to the space–time symmetries. Note that the particle–antiparticle conjugation transformation is an internal transformation. Yet it is tied in with the space–time transformations of spatial reflection and time reversal through *CPT* symmetry. This is a provocative connection and is still only little understood. Refer to Ford (PART, 4), Frisch and Thorndike (PART, 6), Gouiran (PART, 14), Chew, Gell-Mann and Rosenfeld (PART, 2) and Fowler and Samios (PART, 5).

Another kind of symmetry on the astronomical level has to do with black holes. The general theory of relativity predicts that under certain conditions stars will collapse under their own gravitational pull, becoming rapidly smaller and denser, until they reach a stage where their density is so high that space–time in their vicinity is curved to the extreme, causing the stars to 'lose contact with the universe'. No light, other radiation, or matter can escape from such a collapsed star, and an observer approaching too closely will be drawn inward to his destruction. Hence the name 'black hole'. Although these objects cannot be seen, it should be possible to detect them by means of certain effects that they cause. And at least one claim for discovery of such an object has already been made. The only detectable properties of a black hole are its mass, angular momentum and electric charge. Any other characteristic the collapsing star might have had, such as chemical composition or baryon number, becomes lost at the black hole stage. This then is the symmetry we are referring to. There is no detectable difference among the black holes resulting from the gravitational collapse of differing stars, as long as the latter have the same mass, angular momentum and electric charge. Refer to Thorne (REL, 8), Ruffini and Wheeler (REL, 6), Penrose (REL, 12) and Kaufmann (REL, 14).

6
Uses of symmetry in science

'It is because you are a very small animal that you will be
Useful in the adventure before us.'
(A.A. Milne: *Winnie-the-Pooh*)

Now that we have reached some understanding of symmetry — of its general nature, its broad applicability, its terminology — and have become familiar with various kinds of symmetry, both in theory and in nature, we turn to the practical aspects of symmetry and consider its uses. The first use that comes to mind is most likely the esthetic use of symmetry, symmetry in art and design. This is certainly the most important use of symmetry in everyday life and the only one that most people have any familiarity with. But it is just this use of symmetry that we avoid discussing, since I do not feel competent to do so. We only mention that among the fields in which symmetry plays such a role are painting, sculpture, photography, graphic design, industrial design, architecture, music, choreography and poetry. The interested reader is encouraged to make his own investigation of the esthetic use of symmetry. Refer to Weyl (SYM, 1) and listings under ART in the bibliography. We do, however, look into another use of symmetry, its scientific use.

The symmetry principle

There was a moment's silence while everybody thought.
  'I've got a sort of idea,' said Pooh at last, but I don't suppose it's a very good one.'
  'I don't suppose it is either,' said Eeyore.
(A.A. Milne: *The House at Pooh Corner*)

All scientific applications of symmetry are based on the principle that *identical causes produce identical effects*. This seems obvious and hardly worth mentioning; it might be obvious, but it is very worth mentioning, since very useful results follow from the principle. Actually, from a more advanced point of view it is not so obvious. In a more advanced presentation it is preferable to use the word 'equivalent' instead of 'identical', to make the principle broader and more useful. But then there arise questions about which causes are equivalent, which effects are equivalent, and in what way are they equivalent. Sometimes it is not even clear how to distinguish between cause and effect. One is finally led to the recognition that this principle, that equivalent causes produce equivalent effects, is not something imposed upon us by, say, logical necessity. Rather it is a product of our own manufacture; it is the way we tend to interpret our observations of nature. We cannot imagine science existing without such a principle. So we *define* cause, effect and equivalence for each situation in such a way that the principle will hold.

PROBLEM

Compare with the *seemingly* contradictory discussion by R.P. Feynman at the end of chapter 6 of (PHYS, 14). Why is there no contradiction?

But enough of this philosophizing! It is not taking us very far toward practical results. In our introductory approach 'identical' is perfectly satisfactory, and we will have no trouble distinguishing between cause and effect. So identical causes produce identical effects. There is, however, no principle stating that identical effects are produced by identical causes. Identical effects might indeed be brought about by identical causes, but different causes under the proper circumstances can also produce identical effects. For example, a force of magnitude F acting on a body gives the body a certain acceleration. But, because of the parallelogram rule for composition of forces, two forces, each of magnitude $F/\sqrt{2}$ and at right angles to each other, impart to the body the same acceleration as the single force (fig. 6.1). Here we have different causes, different sets of forces acting on the same body, producing identical effects, equal accelerations of the body. Other effects of these two sets of forces might very well not be identical. For instance, they might cause the body to rotate differently in each case.

Now consider the cause and its effect as making up a system. (In the previous example the system would be composed of the body, the set of forces acting on the body, and the resulting acceleration of the body.) And consider transformations of this system. Some transformations might leave the cause unchanged, and some might leave the effect unchanged. Transformations leaving the cause unchanged

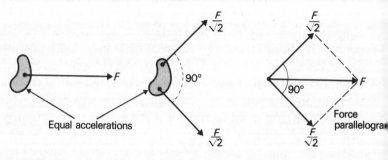

Fig. 6.1
Different causes (sets of forces on the same body) producing identical effects
(accelerations)

must also leave the effect unchanged, because of the principle that identical causes produce identical effects. This set of transformations, under which both the cause and the effect are invariant, is called the *symmetry group of the cause*. Transformations leaving the effect unchanged do not have to leave the cause unchanged, since identical effects can be produced by different causes. So the set of invariance transformations of the effect, the *symmetry group of the effect*, consists of all transformations of the symmetry group of the cause and possibly more. Thus, referring to our discussions of degree of symmetry in chapter 2, we have the very useful result that *the symmetry of the effect is at least that of the cause* or, which is the same, *the effect is at least as symmetric as the cause*. This result is called *the symmetry principle*. Refer to Jaeger (PHYS, 4) for a brief discussion.

Group theory. The symmetry groups of the cause and effect are indeed mathematical groups. A precise formulation of the symmetry principle is: *The symmetry group of the cause is a subgroup of the symmetry group of the effect or coincides with it*.

The symmetry approach

'But if, when in, I decide to practise a slight circular movement from right to left – or perhaps I should say,' he added, as he got into another eddy, 'from left to right, just as it happens to occur to me, it is nobody's business but my own.'
(A.A. Milne: *The House at Pooh Corner*)

The symmetry principle can be directly applied to the solution, partial solution or simplification of problems. Most problems ask about the effect of a given cause, such as the motion of a body under given forces or the currents in a given electric circuit. If a cause is given, then its symmetry can be determined, and this, by the symmetry principle, is at least part of the symmetry of the effect. Such knowledge about the effect is often sufficient for solving the problem fully or partially or at least for simplifying the problem to some degree. The following examples illustrate the application of the symmetry principle.

In our first example we prove that the orbit of a planet about its sun lies completely in a plane and that this plane passes through the center of the sun. Our only assumption is that the sun and planet are each spherically symmetric. We assume nothing about the nature of the force between the two bodies, but we use Isaac Newton's first law, that force causes acceleration. (In its usual formulation, equivalent to this one, the first law of Newton states that in the absence of forces a body remains at rest or moves with constant velocity in a straight line.) Note that this result can also be obtained, as is commonly done in physics textbooks, directly from Newton's second law, without invoking the symmetry principle.

Consider the situation at any instant. The planet has a certain position and a certain instantaneous velocity relative to the sun. Of all the planes passing through the line connecting the centers of the planet and the sun only one is parallel to the direction of the planet's instantaneous velocity. We call this plane the plane of symmetry (fig. 6.2). The cause consists of the sun and the planet with its instantaneous position and velocity. The effect is the acceleration of the planet.

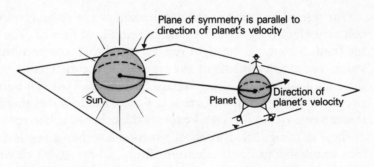

Fig. 6.2
Planet's acceleration, by symmetry principle, is parallel to plane of symmetry

The cause has reflection symmetry with respect to the plane of symmetry (fig. 6.2). The sun and the planet are reflection symmetric with respect to any plane through their centers, since they are spherically symmetric, and the plane of symmetry passes through both centers. The position of the planet is not changed by reflection with respect to the plane of symmetry, again since the plane passes through the planet's center. The direction of the planet's instantaneous velocity is parallel to the plane of symmetry, so this velocity is also invariant under reflection with respect to the plane. Thus the cause is reflection symmetric with respect to the plane of symmetry. The effect, the acceleration of the planet, must then have this symmetry (at least). Therefore the direction of the planet's acceleration must be parallel to the plane of symmetry. (If the acceleration had a component perpendicular to the plane, it and its reflection image would not be in the same direction.) So the planet's velocity, which is parallel to the plane of symmetry, undergoes a change of direction parallel to the plane of symmetry and thus remains parallel to this plane. In this way we see that the planes of symmetry of the system at all instants are in fact one and the same plane and that the motion of the planet is confined to this plane.

PROBLEM

The plane of symmetry is not uniquely defined when the planet is at rest or moving directly toward or away from the sun. Use the symmetry principle to prove that in these cases the orbit of the planet is on a straight line through the center of the sun.

Our second example of the application of the symmetry principle concerns electricity. Consider the DC circuit of fig. 6.3. The two voltage (e.m.f.) sources, the eight resistors, and their connections are the cause. The effect consists of the currents in the different branches of the circuit and the resulting voltages between all pairs of points of the circuit. By G.R. Kirchhoff's first law we arbitrarily designate four of the currents i_1, i_2, i_3, i_4 and express the other four currents in terms of these as in fig. 6.4. Kirchhoff's second law then gives us a set of four simultaneous linear equations for i_1, i_2, i_3, i_4, which we do not show here. These equations can be solved and the solutions used to calculate the other four currents. With all the currents known, the voltage between any pair of points can be calculated. For example, referring to fig. 6.4, we might want to know the voltage between A and C, V_{AC}, the voltage between B and C, V_{BC}, and the voltage between D and E, V_{DE}.

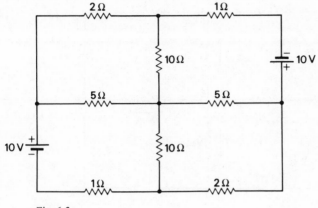

Fig. 6.3
DC circuit

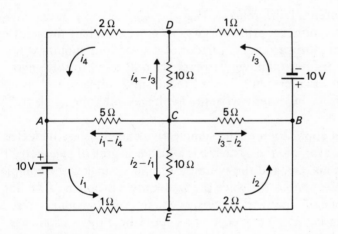

Fig. 6.4
DC circuit with currents indicated

A glance at fig. 6.3 reveals that the cause has symmetry, 2-fold rotational symmetry about the axis through the central junction and perpendicular to the plane of the paper. So the symmetry principle can be invoked, and the effect, the currents and voltages, must be at least as symmetric as this. To help us see what this gives, we rotate the system by 180° as in fig. 6.5. Now compare the rotated system of fig. 6.5 with the unrotated system of fig. 6.4. Symmetry of the effect gives us the relations $i_1 = i_3$ and $i_2 = i_4$, for the currents, and $V_{BC} = V_{AC}$ and $V_{DE} = V_{ED}$, for the voltages. And since, by the nature of

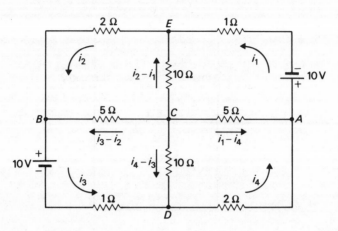

Fig. 6.5
DC circuit rotated by 180°

voltages (potential differences), $V_{DE} = - V_{ED}$, we conclude that $V_{DE} =$ (

Thus the symmetry principle gives us more than half the solution to the problem in this example. Instead of solving four simultaneous equations for four unknown currents, we only need to solve two equations for two currents. One of the desired voltages, V_{DE}, is now known, and of the other two desired voltages only one needs to be calculated.

Our third application of the symmetry principle is again an electrical one. In this example the system has a higher degree of symmetry than in the previous one, and the symmetry principle makes the complete solution quite simple. We solve the well-known problem of finding the resistance of a network of twelve equal resistors connected so that each resistor lies along one edge of a cube, where the resistance is measured between diagonally opposite vertices of the cube, such as between vertices A and H in fig. 6.6. Let r denote the resistance of each resistor. We imagine a voltage V applied between vertices A and H. As a result, current I enters the network at A, branches through the various resistors, and leaves the network at H. The resistance of the network from A to H is then $R = V/I$ by G.S. Ohm's law. The cause is the network and the applied voltage, while the currents in the resistors and the corresponding voltage drops comprise the effect.

We discussed the rotational and plane reflection symmetry of the cube in chapter 3. But the cause in the present example does not possess the full symmetry of the cube in spite of all the resistors being

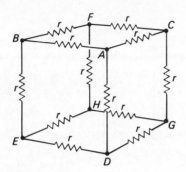

Fig. 6.6
Cube of resistors

equal, since vertices A and H are distinguished from the other vertices and from each other, as the current enters the network at A and leaves at H. So the symmetry transformations of the cause consist of only those symmetry transformations of the cube that do not affect vertices A and H: rotations by $120°$ and $240°$ about the diagonal AH (that is, diagonal AH is an axis of 3-fold rotational symmetry) and reflections with respect to each of the three planes $ABHG$, $ACHE$ and $ADHF$ as in fig. 6.7. Then by the symmetry principle the effect must also have this symmetry. We can use this fact to find the current in each resistor of the network.

Now refer to figs. 6.6 and 6.7. The current I entering the network at vertex A splits among three branches and flows to vertices B, C and D. Due to the 3-fold rotational symmetry the current divides equally, so that current $I/3$ flows in each branch AB, AC and AD as in the diagram of fig. 6.8. The current $I/3$ entering vertex B then divides again between two branches and flows to vertices E and F. It divides equally between the two branches because of the reflection symmetry with respect to plane $ABHG$. Thus current $(I/3)/2 = I/6$ flows in each branch BE and BF. Similar reasoning also gives current $I/6$ in each of the branches CF, CG, DG and DE as in the diagram of fig. 6.9.

After this stage the symmetry takes care of itself. Current $I/6$ enters vertex E from each of vertices B and D, producing current $2(I/6) = I/3$ leaving E and flowing to H. Similarly current $I/3$ enters H from each of vertices F and G. The three currents $I/3$ entering vertex H join to give current $3(I/3) = I$ leaving the network. This is illustrated in the diagram of fig. 6.10.

Now the voltage V between A and H equals the sum of voltage drops

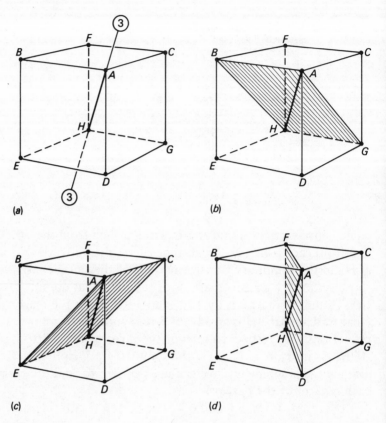

(a)

(b)

(c)

(d)

Fig. 6.7
Symmetry of the cause for resistance of cube of resistors between A and H.
(a) Axis of 3-fold rotational symmetry. (b)–(d) Planes of reflection symmetry

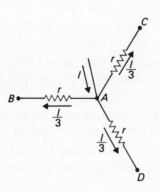

Fig. 6.8
Rotational symmetry requires equal first division of current entering network

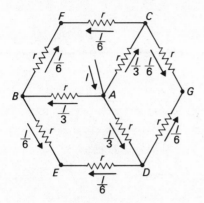

Fig. 6.9
Plane reflection symmetry requires equal second division of current

between these two vertices, where the sum may be computed over any continuous path connecting A and H. Let us take path $ABEH$. By Ohm's law the voltage drop on a resistor is the resistance times the current in the resistor. So, referring to the diagram of fig. 6.10, the voltage drop from A to B is $rI/3$. From B to E the voltage drop is $rI/6$, and from E to H it is $rI/3$. Adding these together we obtain $V = 5rI/6$. Thus the resistance of the network between A and H is $R = V/I = (5rI/6)/I = 5r/6$. This is the solution of the problem.

PROBLEM

Using the symmetry principle, show that vertices B, C and D are equipotential points and so are vertices E, F and G. Prove this also by electrical considerations in the diagram of fig. 6.10.

PROBLEM

Use the symmetry principle to find the resistance of a network of twelve equal resistors connected so that each resistor lies along the edge of an octahedron, where the resistance is measured between a pair of opposite vertices. Although the same number of resistors is involved here as in the cube problem, this problem is simpler. The resistance of the octahedral network is $R = r/2$, where r is the resistance of each resistor.

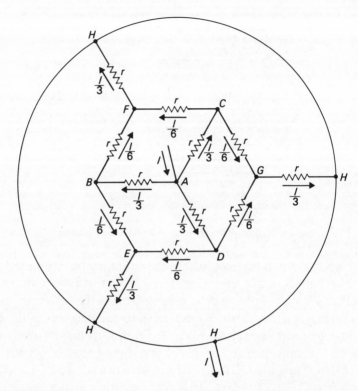

Fig. 6.10
Rejoining of divided currents in two stages before leaving network

These examples illustrate the practical application of the symmetry principle and, through it, the principle that identical causes produce identical effects. The method of application to problems is to keep continual watch for symmetries in the problem and, by means of the symmetry principle, to take full advantage of whatever symmetry exists to simplify or solve the problem. This symmetry approach is very useful and labor-saving, whether in study, in research or in everyday life, and is well worth cultivating.

Symmetry in research

'Without Pooh,' said Rabbit solemnly as he sharpened his
pencil, 'the adventure would be impossible.'
(A.A. Milne: *Winnie-the-Pooh*)

The problems of basic scientific research, however, are usually of the opposite kind. Given the effect, one tries to find the cause. The effect in such a problem is one or more natural phenomena as observed and measured by experimental scientists. The theoretician attempts to explain these phenomena as being produced by some cause. Such an explanation is called a theory. A theory is considered to be better the more different phenomena it explains and the simpler the cause that is supposed to be producing them. But simplicity is not a standardized concept and is largely a matter of taste. (Refer to Gardner (SCI, 7).) Yet it is generally agreed by scientists that symmetry contributes greatly to simplicity. So in devising a theory for a given set of natural phenomena, the theoretician usually tends to assume as symmetric a cause as possible. And just how symmetric can a cause be? Here the symmetry principle works in reverse and tells us that the cause can be no more symmetric than the effect. The theoretician first identifies the symmetry of the phenomena that he wishes to explain. (This symmetry is often far from obvious.) Then he constructs his theory so that the cause will have just this same symmetry, if possible. If this is not possible, he must assume a less symmetric cause and include in his theory an explanation of why the effect is more symmetric than the cause. But he can never assume that the cause has a higher degree of symmetry than the phenomena being explained. This would violate the symmetry principle and thus also the principle that identical causes produce identical effects.

What most often happens, though, is that the symmetry of a set of natural phenomena is only approximate. (Refer to chapter 4.) Then the first step toward a theory is to determine the ideal symmetry which is only approximated by the phenomena. This can be very difficult, if the symmetry is far from exact. To obtain as symmetric a cause as possible the theoretician constructs a theory such that the cause will have broken symmetry with the ideal symmetry of the effect as the exact symmetry limit. Thus the exact symmetry limit produces, through the symmetry principle, the ideal symmetry of the phenomena, while the symmetry breaking brings about the deviation from this ideal symmetry. Sometimes the theory is complete and the symmetry breaking mechanism is included within its framework, and sometimes the cause of the breaking is left as a mystery to be cleared up when more experimental facts are known or a better theory is found.

For an example of such symmetry considerations in scientific research we turn to nuclear physics. The basic problem of nuclear physics is the strong interaction among protons and neutrons, the nuclear force binding these particles together to form nuclei. This nuclear force is not yet completely understood. On the other hand, the electromagnetic force among protons and neutrons is very well understood. These two forces being the cause, the effect consists of such phenomena as the various properties of all kinds of nuclei and the results of scattering experiments, in which protons and neutrons are made to collide. This effect is found to exhibit the following approximate symmetry: two kinds of nuclei differing only in that one of the neutrons in one kind is replaced by a proton in the other have certain similar properties (though their electric charges, for example, are clearly different); and in scattering experiments similar results are obtained whether the interacting particles are two protons, two neutrons, or a proton and a neutron.

So nuclear phenomena are approximately symmetric under interchange of proton and neutron. This symmetry is called charge symmetry, since the major difference between the proton and neutron is their different electric charges. Given that the effect has approximate charge symmetry, theoreticians attempt to incorporate broken charge symmetry into their theories by assuming that the nuclear force is exactly charge symmetric, is completely blind to any difference between protons and neutrons, while the electromagnetic force, which certainly discriminates between the proton (charged) and the neutron (neutral), breaks this symmetry. (In the hypothetical exact symmetry limit of no electromagnetic force, when the symmetry breaking factor is absent, the approximate charge symmetry of nuclear phenomena should become exact.) This assumption has proved to be very successful, and the deviations from exact charge symmetry are indeed well explained by the effect of the electromagnetic force. (Actually, permutation symmetry, which holds when the two particles are identical but not otherwise, has an effect also and must be taken into account. Refer to chapter 5.) See Feynman, Leighton and Sands (PHYS, 13).

More such examples in the field of elementary particle physics are connected with isospin symmetry and SU(3) ('eightfold way') symmetry. Refer to Ford (PART, 4), Frisch and Thorndike (PART, 6), Gouiran (PART, 14), Chew, Gell-Mann and Rosenfeld (PART, 2) and Fowler and Samios (PART, 5).

But even in science we cannot completely avoid the esthetic aspect of symmetry. If we eavesdrop on private discussions among scientists, we might hear expressions such as, 'This is a beautiful theory (of ours)!' or, 'His theory is rather ugly.' Both theories might be equally good, in that they both explain the same natural phenomena equally well.

In fact, the 'ugly' theory might even be better. But the 'beautiful' one attracts more attention.

What makes a theory beautiful? This is, of course, a subjective matter, and in science, too, beauty is in the eye of the beholder. But an opinion poll would reveal that simplicity and symmetry play decisive roles in determining whether a theory appears beautiful or not to most scientists, though simplicity is also subjective and both simplicity and symmetry are relative. Most scientists are prejudiced in favor of (what they consider to be) beautiful theories and feel (albeit irrationally) that nature should be described by beautiful theories. (Even such a nonscientist as the poet John Keats was apparently affected by this feeling, which he expressed in his 'Ode on a Grecian Urn' as, 'Beauty is truth, truth beauty. . .') This is why, as we mentioned above, theories are considered better the simpler (and more symmetric and beautiful) they are. Albert Einstein's general theory of relativity, for example, if far from simple to most physicists, though it is simpler than other theories that might explain the same phenomena equally well. It has a high degree of symmetry and is considered very beautiful by most people familiar with it. For general relativity refer to Bergmann (REL, 1) and Sciama (COS, 15).

This tendency to prefer a beautiful theory to a less beautiful one persists even when the uglier one gives better results! In such cases scientists often stick with the less successful one and follow it up, when they have the feeling that, being the more beautiful theory, it is basically more correct but needs modification. It has happened that scientists abandoned beautiful theories for less beautiful ones that seemed better at the time and lived to regret it. P.A.M. Dirac (SCI, 1) tells of the following case.

Erwin Schrödinger, trying to find a beautiful theory to describe events on the atomic level and working from a mathematical point of view, obtained an equation that was both beautiful and consistent with special relativity. He immediately applied it to the hydrogen atom but got results that were in disagreement with experiment. Then he noticed that a rough approximation to his equation, that ignored the requirements of relativity, gave results that agreed with observation. So he published his approximate equation, which is known today as the Schrödinger equation. Because of his delay the relativistic equation is now known as the Klein–Gordon equation, although it was first discovered by Schrödinger.

What happened is this. It was not known at that time that the

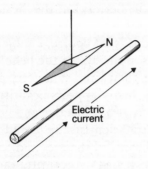

Fig. 6.11
Mach's dilemma

electron has spin. The relativistic equation, though perfectly valid for spinless particles, is simply not applicable to the electron in the hydrogen atom and therefore gave bad results. The nonrelativistic approximation is insensitive to spin and works for the hydrogen atom to fair accuracy.

The correct relativistic equation for the electron was obtained by Dirac himself, although he modestly refrained from mentioning this fact in his article (SCI, 1), and is known, appropriately, as the Dirac equation. This beautiful equation also predicts the existence of another kind of particle, having the same mass as the electron but carrying positive electric charge. Such a particle, now called a positron, had not been discovered when Dirac obtained his equation. The only positively charged particle known then was the proton, and its mass was almost 2000 times too large. He was not discouraged, however, and stuck by his equation until the eventual discovery of the positron proved him right. The moral, according to Dirac, is that 'it is more important to have beauty in one's equations than to have them fit experiment'.

Another piece of symmetry-related scientific gossip is reported by Hermann Weyl (SYM, 1). Ernst Mach received an intellectual shock when he learned as a boy that a magnetic needle, suspended over and parallel to a wire, is deflected in a certain sense, to the left or to the right, when an electric current is passed through the wire in a definite direction (fig. 6.11). The cause here (needle and current) indeed *seems* to possess bilateral symmetry with respect to the plane containing the wire and the needle. So the effect should have the same symmetry, and the needle should not turn. When Mach learned more about the nature of electromagnetism, his dilemma was resolved.

PROBLEM

Can you resolve Mach's dilemma?

In chapter 5 on symmetry in nature we thrice mentioned a connection between invariance of the laws of physics and the conservation of a physical quantity: spatial displacement invariance and conservation of momentum, temporal displacement invariance and conservation of energy, and rotational invariance and conservation of angular momentum. These conservation laws are related to more or less concrete transformations, that can be pictured and can even be intelligibly discussed at the level of this book. But not so with the other, more abstract, conservation laws: conservation of electric charge, baryon number, lepton numbers, isospin, hypercharge, and others. In analogy with the conservation laws for momentum, energy and angular momentum theoreticians would like to find symmetries of physics related to these latter conservation laws as well. But even when this has been done, the transformations involved are of too abstract a nature to be described here. And to complicate matters further, some of these conservation laws are only approximate and should therefore be related to broken symmetries. This is a fascinating field (for me, at least) and is a subject of current research in theoretical physics. Unfortunately, a more detailed discussion of the relation between physical symmetries and conservation laws is beyond the scope of this book. Refer to Feinberg and Goldhaber (CONS, 1), Wigner (CONS, 2), Aharoni (CONS, 3), Feynman, Leighton and Sands (PHYS, 13) and Feynman (PHYS, 14).

Conclusion

'Is that the end of the story?' asked Christopher Robin.
'That's the end of that one. There are others.'
(A.A. Milne: *Winnie-the-Pooh*)

This concludes my attempt to infect you with the symmetry disease. If you have caught the bug, you will be the first to know it, and my efforts will have met with as much success as could be hoped for. If you find yourself attracted to one or more symmetry related fields, say in science, mathematics or art, I will consider this a success as well. And even if this book lit no spark of interest in you, but only contributed, however slightly, to increasing your symmetry consciousness and awareness of symmetry in your environment and life, I would not be disappointed.

Please do not mistake my intention. I am certainly not calling for a symmetry revolution in our lives, nor even for any increase of symmetry. (Clearly, symmetry carried to the extreme would be utterly boring: each to his own taste here.) However, I *do* propose that an increased understanding or even a heightened awareness of symmetry can not only be very useful, but can bring much enjoyment as well. The principles of the use of symmetry in science were presented in chapter 6 along with several examples. The examples were necessarily very elementary, and you must accept my assurance that in more advanced science the role played by symmetry is quite impressive. Group theory then becomes indispensable. For example, two fields (among others) in which symmetry considerations are of major importance are solid state physics (such as the properties of crystals) and quantum mechanics. As to the enjoyment that symmetry can bring, this follows from the general effect that understanding usually deepens and broadens pleasure, though to an individually varying degree. Does not an increased understanding of music, such as through a music appreciation course, heighten one's enjoyment of music? And does not a familiarity with wine lore enhance the pleasure of wine drinking?

I have attempted to open your eyes to the world of symmetry and hope you find it a good world and make the most of it.

Bibliography

Owl ... could spell his own name WOL, and he could spell
Tuesday so that you knew it wasn't Wednesday, and he could
read quite comfortably when you weren't looking over his
shoulder and saying 'Well?' all the time, and he could —
'Well?' said Rabbit.
(A.A. Milne: *The House at Pooh Corner*)

No claim is made for completeness of this bibliography. The entries consist of books, articles and films that I have read, viewed, come across in libraries, read reviews of, or seen references to, and which I consider relevant to the text and not more difficult than a certain vaguely defined level of difficulty. For the purpose of this book the entries are classified according to a rather unsystematic but convenient list of subjects. Each entry was assigned to a main subject and possibly a number of minor subjects and appears under its main subject with its minor subjects, if any, listed in parentheses following the entry. The minor subjects are cross-referenced. Each subject is indicated by a mnemonic abbreviation, and the subjects are arranged in alphabetical order of their mnemonics. References in the text have the form (ART, 5), meaning the fifth entry under the subject 'Symmetry in art', whose mnemonic is ART. The order of listing of entries under each subject is of no significance.

I will be grateful for suggestions of possible additions to this bibliography.

ANTI — Antimatter

1 H. Alfvén: *Worlds–Antiworlds: Antimatter in Cosmology* (W.H. Freeman and Co., 1966). (COS)
2 H. Alfvén: 'Antimatter and Cosmology', *Scientific American*, vol. 216, no. 4, p. 106 (April 1967). (COS)
3 L.C. Hall and C.G. Mena: *Anti-Matter* (Animation Workshop and Physics Department, University of California at Los Angeles, 1973). This is a 12-minute color film with sound.

EXC, 1; PART, 8, 10, 11, 14; PHYS, 10, 13, 14; REF, 1; SCI, 4; TIM, 2, 4.

APROX — Approximate symmetry

PART, 10; PHYS, 9, 13, 14.

ART — Symmetry in art

1   E.B. Edwards: *Pattern and Design with Dynamic Symmetry* (Dover Publications, 1967, originally published 1932).
2   M.C. Escher: *The Graphic Work* (Hawthorn Books, 1967) and *The World of M.C. Escher* (Harry N. Abrams, 1971).
3   J. Hambidge: *The Elements of Dynamic Symmetry* (Dover Publications, 1967, originally published 1926).
4   J. Hambidge: *Practical Applications of Dynamic Symmetry* (Devin-Adair Co., 1965).
5   E.E. Helm: 'The vibrating string of the Pythagoreans', *Scientific American*, vol. 217, no. 6, p. 92 (December 1967).
6   C.H. MacGillavry: *Symmetry Aspects of M.C. Escher's Periodic Drawings* (A. Oosthoek, 1965). (CRYS, COL)
7   J. Schillinger: *The Mathematical Basis of the Arts* (Johnson Reprint, 1948).
8   G.D. Birkhoff: *Aesthetic Measure* (Cambridge, 1933, out of print): 'A mathematical theory of aesthetics and its application to poetry and music', *The Rice Institute Pamphlet*, vol. 19, no. 3 (July 1932); 'A mathematical approach to aesthetics', *Scientia*, vol. 50. p. 133 (1931); and 'Mathematics of aesthetics', reprinted in J.R. Newman: *The World of Mathematics* (Simon and Schuster, 1956), vol. 4, p. 2185.
9   O. Jones: *Grammar of Ornament* (Van Nostrand Reinhold Co., 1972, originally published 1856).
10 G. Kepes (editor): *Module, Proportion, Symmetry, Rhythm* (George Braziller, 1966). (CRYS, LIV, POLY)
11 M. Ghyka: *The Geometry of Art and Life* (Sheed and Ward, 1946). (LIV, POLY)

SYM, 1.

COL — Color symmetry

1   A.V. Shubnikov, N.V. Belov and others: *Colored Symmetry* (Pergamon Press, 1964). (CRYS, GEO, GRP, POLY)
2   A.L. Loeb: *Color and Symmetry* (John Wiley and Sons, 1971). (CRYS, GEO)

CONS — Conservation laws

1   G. Feinberg and M. Goldhaber: 'The conservation laws of physics', *Scientific American*, vol. 209, no. 4, p. 36 (October 1963). (PART)
2   E.P. Wigner: 'Symmetry and conservation laws', *Proceedings of the National Academy of Sciences*, vol. 51, no. 5 (May 1964), reprinted in *Symmetries and Reflections* (PHYS, 6). (PHYS)
3   J. Aharoni: *Lectures on Mechanics* (Oxford University Press, 1972). (REL)

PART, 1, 2, 4, 6, 14; PHYS, 7, 9, 13, 14.

COS — Cosmology, astronomy and astrophysics

1   H. Bondi: *The Universe at Large* (Doubleday and Co., 1960).
2   G. Gamow: *The Creation of the Universe* (Viking Press, 1961).
3   P.W. Hodge: *Galaxies and Cosmology* (McGraw-Hill Book Co., 1966).
4   P.W. Hodge: *Concepts of the Universe* (McGraw-Hill Book Co., 1969).
5   F. Hoyle: *The Nature of the Universe* (New American Library, 1950).
6   F. Hoyle: *Frontiers of Astronomy* (New American Library, 1955).
7   F. Hoyle: *Astronomy* (MacDonald and Co., 1962).
8   B. Lovell: *Our Present Knowledge of the Universe* (Harvard University Press, 1967).

128   *Bibliography*

9   B. Lovell: *The Individual and the Universe* (Oxford University Press, 1961).
10   B. and J. Lovell: *Discovering the Universe* (Benn, 1968).
11   J.P. Ostriker: 'The nature of pulsars', *Scientific American*, vol. 224, no. 1, p. 48 (January 1971). (EXC)
12   P.J.E. Peebles and D.T. Wilkinson: 'The primeval fireball', *Scientific American*, vol. 216, no. 6, p. 28 (June 1967).
13   M.J. Rees and J. Silk: 'The origin of galaxies', *Scientific American*, vol. 222, no. 6, p. 26 (June 1970).
14   M.A. Ruderman: 'Solid stars', *Scientific American*, vol. 224, no. 2, p. 24 (February 1971). (EXC)
15   D.W. Sciama: *The Unity of the Universe* (Faber and Faber, 1959). (REL)
16   J. Singh: *Great Ideas and Theories of Modern Cosmology* (Dover Publications, 1961).
17   H.C. Arp: 'The evolution of galaxies', *Scientific American*, vol. 208, no. 1, p. 70 (January 1963).
18   G. Abell: *Exploration of the Universe* (Holt, Reinhart and Winston, 1969).
19   W. Bonnor: *The Mystery of the Expanding Universe* (Macmillan Co., 1964).
20   H. Bondi: *Cosmology* (Cambridge University Press, 1961).
21   M.M. Nieto: *The Titius–Bode Law of Planetary Distances: Its History and Theory* (Pergamon Press, 1972).

ANTI, 1, 2; REL, 1, 3, 6, 8, 12, 14; SCI, 3, 4.

CRYS — Crystals and crystallography

1   W.A. Bentley and W.J. Humphreys: *Snow Crystals* (Dover Publications, 1962, originally published 1931).
2   A. Holden and P. Singer: *Crystals and Crystal Growing* (Doubleday and Co., 1960). (GEO)
3   M.A. Jaswon: *Mathematical Crystallography* (Longmans, Green and Co., 1965). (GEO)
4   N. Mott: 'The solid state', *Scientific American*, vol. 217, no. 3, p. 80 (September 1967). (EXC)
5   A.F. Wells: *The Third Dimension in Chemistry* (Oxford University Press, 1956). (POLY, LIV)
6   A. Holden: *The Nature of Solids* (Columbia University Press, 1965).
7   C. Bunn: *Crystals: Their Role in Nature and in Science* (Academic Press, 1964). (GEO)

ART, 6, 10; COL, 1, 2; GEO, 1; LIV, 8; PHYS, 1, 4, 11, 12; SYM, 1.

EXC — Pauli exclusion principle

1   G. Gamow: 'The exclusion principle', *Scientific American*, vol. 201, no. 1, p. 74 (July 1959). (ANTI, PART)

COS, 11, 14; CRYS, 4; SCI, 4.

GEO — Geometric symmetry

1   H.S.M. Coxeter: *Introduction to Geometry* (John Wiley and Sons, 1969). (GRP, POLY, CRYS, LIV)
2   C.F. Linn: *A Classroom Guide to Symmetry* (Doubleday and Co., 1968). This is a teacher's guide to (GEO, 3). (GRP)
3   A.G. Razzell and K.G.O. Watts: *Symmetry* (Doubleday and Co., 1968). For children.
4   I.M. Yaglom: *Geometric Transformations* (Random House, 1962).
5   National Council of Teachers of Mathematics (U.S.): *Symmetry, Congruence and Similarity* (1969).

6   W.W. Sawyer: *Prelude to Mathematics* (Penguin Books, 1955).
7   M. and H. Sitomer: *What is Symmetry?* (A. and C. Black, 1970). For children.
8   L. Glasser: 'Teaching symmetry', *Journal of Chemical Education*, vol. 44, no. 9, p. 502 (September 1967).
9   I. Bernal, W.C. Hamilton and J.S. Ricci: *Symmetry: A Stereoscopic Guide for Chemists* (W.H. Freeman and Co., 1972). (PHYS)
10 P. Stapp, J. Bregman, R. Davisson and A. Holden: *Symmetry* (Contemporary Films/McGraw-Hill, 1967). This is a $10\frac{1}{2}$-minute color film with sound.
11 H. Liebeck: *Algebra for Scientists and Engineers* (John Wiley and Sons, 1969). (GRP)
12 S. Schuster and W.O.J. Moser: *Isometries* (College Geometry Project, University of Minnesota). This is a 26-minute color film with sound.
13 H.S.M. Coxeter: *Dihedral Kaleidoscopes* (College Geometry Project, University of Minnesota). This is a 13-minute color film with sound.

COL, 1, 2; CRYS, 2, 3, 7; GRP, 1, 3, 4, 6, 7; PHYS, 4, 11, 12; SYM, 1.

GRP – Group theory

1   P.S. Alexandroff: *An Introduction to the Theory of Groups* (Hafner Publishing Co., 1959). (GEO)
2   G.P. Beaumont and P. Caldwell: *An Introduction to Groups* (Association of Teachers of Mathematics (England), 1964).
3   A.W. Bell: *Algebraic Structures – Some Aspects of Group Structure* (George Allen and Unwin, 1966). (GEO)
4   A. Bell and T. Fletcher: *Symmetry Groups* (Association of Teachers of Mathematics (England), 1970). (GEO)
5   M.E. Munroe: *Ideas in Mathematics* (Addison-Wesley Publishing Co., 1968). (GEO)
6   I. Grossman and W. Magnus: *Groups and Their Graphs* (Random House, 1964). (GEO)
7   F.J. Budden: *The Fascination of Groups* (Cambridge University Press, 1972). I especially recommend this book as an introduction to group theory. (GEO)

COL, 1; GEO, 1, 2, 11; PHYS, 12; SYM, 1.

LIV – Symmetry in living things

1   A.H. Church: *The Relation of Phyllotaxis to Mechanical Laws* (Williams and Norgate, 1916, out of print).
2   A.H. Church: *On the Interpretation of Phenomena of Phyllotaxis* (Hafner Publishing Co., 1968, originally published 1920).
3   M.C. Corballis and I.L. Beale: 'On telling left from right', *Scientific American*, vol. 224, no. 3, p. 96 (March 1971). (REF)
4   M.S. Gazzaniga: 'The split brain in man', *Scientific American*, vol. 217, no. 2, p. 24 (August 1967).
5   R. Snow: 'Problems of phyllotaxis and leaf determination', *Endeavour*, vol. 14, no. 56, p. 190 (October 1955).
6   R.W. Sperry: 'The great cerebral commissure', *Scientific American*, vol. 210, no. 1, p. 42 (January 1964).
7   D'A.W. Thompson: *On Growth and Form* (Cambridge University Press, 1969, originally published 1917).
8   R.W. Horne: 'The structure of viruses', *Scientific American*, vol. 208, no. 1, p. 48 (January 1963). (CRYS, POLY)
9   E. Samuel: *Order: In Life* (Prentice Hall, 1972).
10 D. Kimura: 'The asymmetry of the human brain', *Scientific American*, vol. 228, no. 3, p. 70 (March 1973).

ART, 10, 11; CRYS, 5; GEO, 1; PHYS, 4; SYM, 1; TIM, 5, 7, 10, 11, 12.

PART — Elementary particles

1   V.D. Barger and D.B. Cline: 'High energy scattering', *Scientific American*, vol. 217, no. 6, p. 76 (December 1967). (CONS)

2   G.F. Chew, M. Gell-Mann and A.H. Rosenfeld: 'Strongly interacting particles', *Scientific American*, vol. 210, no. 2, p. 74 (February 1964). (CONS)

3   K.W. Ford: 'Magnetic monopoles', *Scientific American*, vol. 209, no. 6, p. 122 (December 1963).

4   K.W. Ford: *The World of Elementary Particles* (Ginn and Co., 1963). (CONS)

5   W.B. Fowler and N.P. Samios: 'The omega-minus experiment', *Scientific American*, vol. 211, no. 4, p. 36 (October 1964).

6   D.H. Frisch and A.M. Thorndike: *Elementary Particles* (Van Nostrand Reinhold Co., 1964). (CONS)

7   L.M. Lederman: 'The two-neutrino experiment', *Scientific American*, vol. 208, no. 3, p. 60 (March 1963).

8   T.D. Lee: 'Space inversion, time reversal and particle–antiparticle conjugation', *Physics Today*, vol. 19, no. 3, p. 23 (March 1966). (REF, TIM, ANTI)

9   S.B. Trieman: 'The weak interaction', *Scientific American*, vol. 200, no. 3, p. 72 (March 1959).

10 E.P. Wigner: 'Violations of symmetry in physics', *Scientific American*, vol. 213, no. 6, p. 28 (December 1965). (REF, TIM, ANTI, APROX)

11 C.N. Yang: *Elementary Particles* (Princeton University Press, 1962). (REF, ANTI)

12 R.D. Hill: 'Resonance particles', *Scientific American*, vol. 208, no. 1, p. 38 (January 1963).

13 V.W. Hughes: 'The muonium atom', *Scientific American*, vol. 214, no. 4, p. 93 (April 1966).

14 R. Gouiran: *Particles and Accelerators* (Weidenfeld and Nicolson, 1967). (ANTI, REF, TIM, CONS)

CONS, 1; EXC, 1; PHYS, 2; SCI, 4; TIM, 4.

PHYS — Symmetry in physics (and chemistry)

1   P.B. Dorain: *Symmetry in Inorganic Chemistry* (Addison-Wesley Publishing Co., 1965). (CRYS)

2   F.J. Dyson: 'Mathematics in the physical sciences', *Scientific American*, vol. 211, no. 3, p. 128 (September 1964). (PART)

3   L. Pauling and R. Hayward: *The Architecture of Molecules* (W.H. Freeman and Co., 1964).

4   F.M. Jaeger: *Lectures on the Principle of Symmetry and Its Applications in All the Natural Sciences* (Elsevier Publishing Co., 1920, out of print). (GEO, CRYS, LIV)

5   J. Kepler: *The Six-Cornered Snowflake* (Oxford University Press, 1966, originally published 1611).

6   E.P. Wigner: *Symmetries and Reflections* (Indiana University Press and M.I.T. Press, 1967). This is a collection of articles, some of which are listed separately in this bibliography.

7   E.P. Wigner: 'Invariance in physical theory', *Proceedings of the American Philosophical Society*, vol. 93, no. 7 (December 1949), reprinted in *Symmetries and Reflections* (PHYS, 6). (CONS)

8   E.P. Wigner: 'The role of invariance principles in natural philosophy', *Proceedings of the International School of Physics 'Enrico Fermi'*, vol. 29, p. 40 (1964), reprinted in *Symmetries and Reflections* (PHYS, 6).

9   E.P. Wigner: 'Events, laws of nature, and invariance principles', in *The Nobel Prize Lectures* (Elsevier Publishing Co., 1964), reprinted in *Symmetries and Reflections* (PHYS, 6). (CONS, APROX)

10 E.P. Wigner 'Relativistic invariance and quantum phenomena', *Reviews of Modern Physics*, vol. 29, no. 3 (July 1957), reprinted in *Symmetries and Reflections* (PHYS, 6). (REL, REF, ANTI)

11   H.H. Jaffé and M. Orchin: *Symmetry in Chemistry* (John Wiley and Sons, 1965). (GEO, CRYS)
12   A. Nussbaum: *Applied Group Theory for Chemists, Physicists and Engineers* (Prentice-Hall, 1971). (GEO, GRP, CRYS)
13   R.P. Feynman, R.B. Leighton and M. Sands: *The Feynman Lectures on Physics* (Addison-Wesley Publishing Co., 1963). (ANTI, APROX, CONS, REF, SCI, TIM)
14   R.P. Feynman: *The Character of Physical Law* (M.I.T. Press, 1965). (ANTI, APROX, CONS, REF, SCI, TIM)

CONS, 2; GEO, 9; REF, 2.

POLY — Polygons and polyhedrons

1   M.C. Hartley: *Patterns of Polyhedrons* (Edwards Bros., 1951, out of print).
2   H.M. Cundy and A.P. Rollett: *Mathematical Models* (Oxford University Press, 1961).
3   H. Steinhaus: *Mathematical Snapshots* (Oxford University Press, 1969).
4   T.S. Row: *Geometric Exercises in Paper Folding* (Dover Publications, 1966, originally published late nineteenth century).
5   A. Holden: *Shapes, Space, and Symmetry* (Columbia University Press, 1971).
6   M. Gardner: 'Mathematical games', *Scientific American*, vol. 224, no. 3, p. 204 (September 1971).
7   K. Critchlow: *Order in Space* (The Viking Press, 1969).
8   M.J. Wenninger: *Polyhedron Models* (Cambridge University Press, 1971).
9   M.J. Wenninger: *Polyhedron Models for the Classroom* (National Council of Teachers of Mathematics (U.S.), 1966).
10   H.S.M. Coxeter and W.O.J. Moser: *Symmetries of the Cube* (College Geometry Project, University of Minnesota). This is a $13\frac{1}{2}$-minute color film with sound.

ART, 10, 11; COL, 1; CRYS, 5; GEO, 1; LIV, 8.

REF — Reflection

1   M. Gardner: *The Ambidextrous Universe: Left, Right and the Fall of Parity* (New American Library, 1969). (ANTI)
2   V. Fritsch: *Left and Right in Science and Life* (Barrie and Rockliff, 1968). (PHYS)

LIV, 3; PART, 8, 10, 11, 14; PHYS, 10, 13, 14; TIM, 2, 4.

REL — Relativity

1   P.G. Bergmann: *The Riddle of Gravitation* (Charles Scribner's Sons, 1968). (COS)
2   H. Bondi: *Relativity and Common Sense* (Doubleday and Co., 1964).
3   M. Gardner: *Relativity for the Million* (Pocket Books, 1962). (COS)
4   C.V. Durell: *Readable Relativity* (Harper and Row Publishers).
5   L.D. Landau and G.W. Rumer: *What Is Relativity?* (Fawcett World Library, 1972).
6   R. Ruffini and J.A. Wheeler: 'Introducing the black hole', *Physics Today*, vol. 24, no. 1, p. 30 (January 1971). (COS)
7   E.F. Taylor and J.A. Wheeler: *Spacetime Physics* (W.H. Freeman and Co., 1966).
8   K.S. Thorne: 'Gravitational collapse', *Scientific American*, vol. 217, no. 5, p. 88 (November 1967). (COS)
9   A. Einstein: *Relativity: The Special and the General Theory* (Methuen and Co., 1968, originally published 1920).
10   B. Russell: *The ABC of Relativity* (New American Library, 1969).
11   J.L. Synge: *Talking About Relativity* (North-Holland Publishing Co., 1970).
12   R. Penrose: 'Black holes', *Scientific American*, vol. 226, no. 5, p. 38 (May 1972). (COS)

132   Bibliography

13   L. Marder: *Time and the Space-Traveller* (George Allen and Unwin, 1971).
14   W.J. Kaufmann, III: *Relativity and Cosmology* (Harper and Row, 1973). (COS)

CONS, 3; COS, 15; PHYS, 10; SCI, 3, 4; TIM, 6.

SCI — General physical science

1   P.A.M. Dirac: 'The evolution of the physicist's picture of nature', *Scientific American*, vol. 208, no. 5, p. 45 (May 1963).
2   G. Feinberg: 'Ordinary matter', *Scientific American*, vol. 216, no. 5, p. 126 (May 1967).
3   G. Gamow: *One, Two, Three — Infinity: Facts and Speculations of Science* (Viking Press, 1963). (REL, COS)
4   G. Gamow: *Matter, Earth, and Sky* (Prentice-Hall, 1965). (REL, COS, PART, ANTI, EXC)
5   G. Gamow: The series of Mr Tompkins stories, such as *Mr Tompkins in Paperback* (Cambridge University Press, 1967).
6   H.S.W. Massey and A.R. Quinton: *Basic Laws of Matter* (Herald Books, 1965).
7   M. Gardner: 'Mathematical games', *Scientific American*, vol. 221, no. 2, p. 118 (August 1969).
8   R.E. Lapp and Time—Life editors: *Matter* (Time—Life Books, 1969).
9   M. Wilson and Time—Life editors: *Energy* (Time—Life Books, 1969).

PHYS, 13, 14.

SYM — Symmetry in general

1   H. Weyl: *Symmetry* (Princeton University Press, 1952). This is the modern classic on symmetry. Part of this book is reprinted in J.R. Newman: *The World of Mathematics* (Simon and Schuster, 1956), vol. 1, p. 671. (GEO, ART, CRYS, GRP, LIV)

TIM — Time

1   W. Ehrenberg: 'Maxwell's demon', *Scientific American*, vol. 217 no. 5, p. 103 (November 1967).
2   M. Gardner: 'Can time go backward?', *Scientific American*, vol. 216, no. 1, p. 98 (January 1967). (REF, ANTI)
3   S.A. Goudsmit, R. Claiborne and Time—Life editors: *Time* (Time—Life Books, 1966).
4   O.E. Overseth: 'Experiments in time reversal', *Scientific American*, vol. 221, no. 4, p. 88 (October 1960). (PART, REF, ANTI)
5   E.T. Pengelley and S.J. Asmundson: 'Annual biological clocks', *Scientific American*, vol. 224, no. 4, p. 72 (April 1971). (LIV)
6   R. Schlegel: *Time and the Physical World* (Dover Publications, 1969). (REL)
7   A. Sollberger: *Biological Rhythm Research* (American Elsevier Publishing Co., 1965). (LIV)
8   A. Copland: *What to Listen for in Music* (New American Library, 1964).
9   A. Mann: *The Study of Fugue* (W.W. Norton and Co., 1966).
10   G.G. Luce: *Biological Rhythms in Human and Animal Physiology* (Dover Publications, 1970). (LIV)
11   G.G. Luce: *Body Time: Physiological Rhythms and Social Stress* (Pantheon Books, 1971). (LIV)
12   H. Strughold: *Your Body Clock* (Charles Scribner's Sons, 1971). (LIV)
13   H.A. and H.E. Bent: *You Can't Go Back* (Elementary Penguin Productions). This is a 6-minute color film with sound.

PART, 8, 10, 14; PHYS, 13, 14.

Index

A page number in bold type indicates the location of a definition of the corresponding entry